EYES IN THE SOLES OF MY FEET

EYES IN THE SOLES OF MY FEET

From Horseshoe Crabs to Sycamores, Exploring Hidden Connections to the Natural World

CAROLINE SUTTON

Schaffner Press

Schaffner Press, Inc.

This is the First Hardcover Edition, published by Schaffner Press, Inc.
POB 41567, Tucson, Az 85717
Printed and manufactured in the
United States of America

Cover and Interior Design by:
Jordan Wannemacher

Library of Congress Control Number: 2025933411

ISBN:9781639640812
EPUB: 9781639640829
EPDF ISBN: 9781639640829

For Ella & Florence, Arthur & Alice

Contents

Preface

We still don't know how consciousness occurs, although it involves
an enormous number of intracellular properties.

MELANIE CHALLENGER
HOW TO BE ANIMAL

W HEN MY SON'S DOG leapt at a plum-sized mammal perched
on the back of a lawn chair one afternoon, I assumed it was a
mouse. Later my husband Brian asked if I was sure, and I began to wonder.
I knew that voles were devouring the roots of our shrubs, but I'd never seen
one and knew nothing about them other than Brian's distress about the
damage they were doing. When I went online to determine if what I'd seen
was in fact a vole, the first page contained nothing but methods for its total
annihilation. I was shocked. What ethos—and often ignorance—drives our
decisions to nurture or to kill? Soon after, I was tugging weeds with little
white flowers from my yard and began wondering why I should prefer the
surrounding homogeneous grass. Weeds are generally defined as a plant
"growing in the wrong place," which blatantly exposes such categorization
as arbitrary, subjective, and anthropogenic. Again I questioned the cultural,
economic, and aesthetic reasons underlying our choices.

I sharpened this focus when for several years I witnessed the natural
world not only with my own eyes but through the eyes of my granddaughter,

Ella. In March 2020, at the outbreak of the COVID virus, my daughter Sophia and her family fled Brooklyn to live with Brian and me at our house on Long Island near a beach on Peconic Bay. Hour to hour, day to day care of the family during the pandemic became my priority and with it an unforeseen plus: Ella's curiosity enhanced the variety of what we explored together and altered the tempo. From her infancy to age four I saw her curiosity about and burgeoning fears of the plants and animals she came across. I realized that if we are to topple conventional prejudices and aesthetic judgments about the natural world, we can begin as soon as a child comes across a worm or spots a swarm of ants on the kitchen floor and stoops to watch them work, or chooses a clam shell tattooed by a boring sponge over one we find unblemished. Ella taught me to see the beauty in dead flowers I trim in my garden. I taught her that seaweed needs water to survive and witnessed her at the age of two toddle to the water's edge, fling a piece of dead man's fingers into the water and command, "Live." We developed a reciprocal relationship as we wandered the shoreline, expressed in its most literal sense by her comment, "I see the tiny creatures at my feet, you see the sky and the birds."

The idea of considering the many ways to see and otherwise sense the world ignited my curiosity as to how nonhuman species go about their lives using their sensory powers. What I learned over six years researching this book quickly leveled any hierarchical values traditionally accorded human and nonhuman perception. I find it humbling that many crustaceans, echinoderms, gastropods, coelenterates, and at least 1,500 species of fish have the capacity to create light, making it the most common language on the planet, yet one we don't speak. During a full moon, a female Bermuda fireworm might rise from the ocean floor and light up as it moves in circles to attract a male. A vampire squid can eject a radiant cloud of mucus to disorient a predator, and some deep-sea dragonfish have evolved to emit red light, which potential prey cannot see. The language of bioluminescence is supple, filled with possibilities, and not reducible to our own.

As I came across commonplace and extraordinary natural phenomena, I realized that distinctions between the two often blur. Portuguese men-of-war strew the beaches of Florida, for example; beachgoers walk past without a glance, and yet these creatures are a strange miracle of biology: a single man-of-war comprises four discrete, specialized individuals that may or may not even communicate with each other. This configuration led me to think about parallels in human existence, including the institution of marriage.

Many of these essays also illustrate analogy and connection through their form, interweaving as they do human and nonhuman experiences of motherhood, birth, and death. They reveal implicitly that we are not as severed from the natural world as we have endeavored to be over the centuries, nor is our way of interpreting and experiencing such archetypal milestones the only way. Early one morning I watched a loggerhead turtle get up from her crater in the sand and waddle down the beach. The eggs she left behind might well be devoured by a coyote or skunk. In a few months any surviving hatchlings would appear and aim for a glint of starlight on the water, scrambling for a mirage of safety where a barracuda might find it, after all. As a baby turtle launches into the water alone, with no definition of courage, I think about rocking my infant children, feeling their fingers around one of mine, holding their hands as they take first steps. But despite such vigilance, to what will my children be vulnerable?

While researching, I was surprised—and pleased—at the many times I came across a note, often buried far into an article, that scientists know little about such and such a species, or we don't know why such and such is so. We're not really sure why sycamores shed bark, or what induces prairie voles to bond for life, or what bowhead whales communicate to one another with their improvisatory calls. In this information age I had come to believe that just about everything had been explored by now, the wilderness having shrunk and scientific research becoming ever more sophisticated.

Early in the 20th century, for instance, we didn't know that horseshoe crab blood contains proteins that we now use to make our medicines safe. We slaughtered millions of these ancient creatures and ground them up for pig feed. But the story is not over. Perhaps we still have not uncovered the ramifications of horseshoe crabs having ten eyes and an ability to perceive ultraviolet. Our presumption to know the sea is undermined by the fact that what we see must be entirely different from what they see. As we look at these and other ancient species, we are acknowledging that we have no definitive idea where and when consciousness arose. Believing that this is a faculty belonging only to humans has been challenged, not only in studies of mammals like monkeys so similar to ourselves, but also in early cephalopods—octopuses and cuttlefish—with inconclusive results.

Still, what we have been able to determine about ways nonhuman species perceive the world is extensive and profound. I've found it stunning to consider that I will never experience what many species are capable of—to see UV as it streaks across a bay, or to sense the earth's magnetic poles with confounding precision—and yet I'm enriched by knowing these faculties exist. Recognizing our own limitations can lead us to question our assumptions. It can inspire empathy and respect for the plants and animals that share the air we breathe, whose survival is intertwined with ours. Clearly, a fundamental change in our attitudes toward the natural world, not just a quick nod of appreciation, is critical if we are to preserve our wealth of ecosystems for the grandchildren of our grandchildren and beyond.

But without shooting rabbits, how was one to keep them down?
she wondered. It might be a rabbit; it might be a mole. Some creature
anyhow was ruining her Evening Primroses. And looking up, she saw above
the thin trees the first pulse of the full-throbbing star, and wanted to make
her husband look at it; for the sight gave her such keen pleasure.

VIRGINIA WOOLF
TO THE LIGHTHOUSE

When I catch a fish, perhaps I need to have been caught by it.

TIMOTHY MORTON
BEING ECOLOGICAL

WATER

Eyes in the Soles of My Feet

WHEN I FIND A horseshoe crab on the rocky beaches of Peconic Bay, I never know if it's alive or dead. I walk past it as I would ravaged whelks and pebbles and sea glass and clamshells and seaweed. There it lies, Achilles' shield, a rock, an oversized bug; there it lies, turning metal hot with the sun, cold and invisible on a starless night long after the living thing inside has died. I kick over a shell, dappled with parasitic flatworms and slipper shells, and find tidy rows of triangular gills like so many delicate shoulder blades.

This crab doesn't scuttle along the shores of silent seas because it isn't, technically, a crab. It's more closely related to scorpions and spiders. Often referred to as a fossil, which I think of as a rock with the impression of a formerly living thing, it has a phenomenal capacity to survive. The horseshoe crab is virtually the same as it was nearly half a billion years ago; it existed before dinosaurs, and endured as they died off. One factor in its survival may be its blitzkrieg defense system, whose arsenal resides in the blood. Instead of hemoglobin to carry oxygen, this blood has hemocyanin, which sounds strangely lethal, but simply means it contains copper, which lends it color the way a copper drainpipe turns blue-ish green. The critical element, though, is *Limulus* amebocyte lysate (LAL), which detects bacterial endotoxins even in concentrations of one to a trillion, and when that happens LAL bolts into action, surrounding, choking, imprisoning the bacteria in goo for as long as a few weeks. The crabs don't have circulatory systems as we do. Any crack in the exoskeleton opens the way for bacteria to enter and pervade, but at that point the blood releases granules of LAL, which leave the bacteria in a quagmire, neutralized.

Not surprisingly, humans have gotten in on the action. Five laboratories on the East Coast harvest the crabs, alive, strap them to a metal bar, pierce the tissue near the heart and drain about a third of their pale blue blood before releasing them. When medical companies develop new drugs and prosthetic devices that come into contact with our blood, they test samples for bacterial infection by subjecting them to the horseshoe crab's coagulant. In short, any American who has ever gotten an injection is safer because of the horseshoe crab.

Bleeding proponents claim most bled crabs do just fine once returned to the sea; some say only 10% of the crabs die, but tracking the crabs is impossible and the numbers are murky. Environmentalists reveal much higher mortality in the short term, and further, more nebulous loss in a nebulous future since females, weakened by blood deficit, may not climb onto shore at the optimal time to lay eggs, or may not get there at all. In recent years their status has shifted from Near Threatened to Vulnerable—at high risk of extinction in the wild.

When the tide was out recently, I found a huge one far up the beach, its shell dark as a blacksmith's forge and coated in seaweed and slipper shells. I picked it up gingerly, stunned by the weight and unsure if it was alive, and if it were, if its pincers could reach my fingers. As I held it up to the sunlight, two legs moved ever so slightly and slowly. I had disinterred a corpse bent on resuming life in some singular dimension of time—and then the tail, this stiff sharp spear, rose perpendicular to its body—was it more alarmed than I? I walked to the shoreline where I set it down, right side up, and looked for its eyes, tried to look into its eyes.

Later I learned the crab could have righted itself. The spear is not a weapon but ballast with which the crab flips itself over in the sand. And those lightless bumps on the prosoma, or head region, which I could scarcely detect, are compound eyes with 1,000 surfaces like those of a fly that see hundreds of images at once. They look like a honeycomb, and each eye has rods and cones a hundred times the size of ours. The nocturnal crab has five more eyes on the carapace, which I totally missed, and two on the ventral side near the mouth, which may help orient the crab while swimming. Horseshoe crabs see visible light and ultraviolet in sun and moonlight, which keeps it in tune with lunar cycles, so the female knows when to clamber up onto the beach, nestle in the sand and lay eggs for gang fertilization. Add to this: light sensors along the tail synchronize the animal's brain with shifts in darkness and light. It sounds like being a horseshoe crab is a veritable light show—streaks of ultraviolet intersecting with spectral blues, reds, greens, all charged with information about spawning and feeding on crustaceans and mollusks on the ocean floor—and yet there it lies, looking scarcely animate, a bit of alien mystery washed up on shore by an ancient sea when pterodactyls not seagulls winged overhead. But I've read this survivor has lousy vision; it's not 20/20, which is perfection as humans define it.

Years ago, when I gave birth on the night of a full moon in early May, I didn't think about a hefty horseshoe crab on the beach laying 2,000 eggs, about smaller males crowding around for a chance at even anonymous paternity, about sandpipers pecking at the feast. "The maternity ward is full!" reported my midwife who darted from room to room. Two weeks late, my baby rode high in the womb, resisting all medical anxieties and predictions, and I was scheduled for a C-section the following morning. But the tug of the moon must have rippled through me unawares, sending me into labor (as it did so many other mothers, other babies), and now I wonder if horseshoe crabs that night were more conscious than I of the timing to give birth though consciousness sounds ludicrous when associated with a crab. Then again, how often do I operate on sensory or intuitive impulses that have not quite nosed the surface of consciousness and either delude myself that they have or admit to having no idea why or if I did what I did?

I ended a three-year relationship because a certain smell turned me off. Suddenly. Subliminally. Irrevocably. It didn't even feel like my decision.

I've sat on temple floors and church pews and stared at multiple incarnations of Buddha, the graphic martyrdom of Saint Sebastian, the tender astonishment of Mary in Da Vinci's *Annunciation*, all with an inclination to be transported, all with the indefinable sensation that mind exists distinctly from matter, when a rational rejection of things intangible and divine slams like a ceiling attic door with Bessler stairs pulled up and out of sight.

I'm left with the weight of a horseshoe crab in my palms, a handful of sand, my own finite being, all this physical matter, one human body comprised of pretty much the same elements as another. Yet, if I prick my finger, and yours, I know what it feels like to me but can never know what it feels like to you. And what of death, or love? That even shared experience of the world is intricately dissimilar at the very least suggests the existence of *something* not reducible to the darting about of atoms. I find it astounding and strangely comforting that over the course of evolution, we have no idea when consciousness arose. We know that simple things gave rise to more complex ones, but since we can't pinpoint a moment, and can't argue physically that something appeared out of nothing, then we have to assume continuous development. In that case, we today with all our sophisticated technologies share something quite mysterious with an amoeba.

At the age of four, my daughter (who happened to be very conscious of her own being as female) asked, "Does a dog know it's a dog? Does a dog know another dog is a dog?"

"Good question!" I stalled. Should I tell her, no, a dog does not know it's "a domesticated carnivorous mammal that typically has a long snout, an acute sense of smell, and a barking, howling, or whining voice" as the dictionary informs us. Nor does a dog think it's "a person regarded as unpleasant, contemptible, or wicked," which writers of the dictionary have inferred is the connotative meaning. But how can I be sure? My dog used to go to the door and scratch so my husband would get up from the most comfortable chair in the room to open it, whereupon the dog would wheel around and dart to the chair, curl up, and snooze. So although I doubt the dog applied a moral judgment to its actions—sneaky, manipulative, Darwinian—it knew it liked the chair as much as my husband did, which gives them something essentially in common that might blur the line even infinitesimally between dog-ness and human-ness. And by generally accepted definition, we are conscious beings. Can consciousness of comfort and consciousness of self be mutually exclusive?

What does a horseshoe crab sense when a medical technician strings it up like a prisoner of war and leaves it hanging for indeterminate, non-self-determined, periods of time? What sensations dart through its blue blood before it drips, drips, drips into a glass jar? Does it sink into a dormant state—numb, near paralyzed—as people do when traumatized? A *New Yorker* article reported that Syrian teenagers whose lives were decimated by war have fallen into coma-like sleep that lasts for months, even years, in the face of hopeless lack of control.

In a rational world, I don't know why I find the photos of bleeding crabs so disturbing. Were they not so valued for their blood, fisheries might slaughter even more crabs for use as bait to catch eels and whelk, which is currently allowed in some states. In the twentieth century populations declined at a feverish pace when horseshoe crabs were steamed, chopped up, ground into meal and spread as fertilizer or fed to hogs. Were they not so prized, pharmaceutical companies would still be testing rabbits, which horrifies us more since they look picturesque nibbling clover on a lawn and watching us with their warm brown eyes. Perhaps it's the collision of eons marking the horseshoe crab's history and our medical needs for tomorrow, the stark eradication of millions of years hurling themselves on the shore only to recede limply back to sea that strikes me as sacrilegious.

And, more immediately, blind. When I see rows and rows of strung up crabs, I now know many were tossed onto fishing boats by their tails, which are vital to their survival, that many were kept out of water for hours before the bleeding. The pandemic and the race for a vaccine created more critical stress on populations with East Coast labs harvesting 700,000 crabs in 2021 and twice that number in 2023—despite the creation years before of a synthetic replacement for LAL. Recombinant Bacterial Endotoxins Tests (rBET) mimic the protein that detects endotoxins, and some scientists claim they're more consistent than LAL. Acceptance was delayed for years by sluggish steps and internal debate at US Pharmacopeia, which oversees safety standards of new medical products. Finally, in July 2024, the organization announced its approval, an overdue but promising move for the plight of the crabs.

Mere decades have passed in my daughter's life, but now she has grown up and is ready to have a baby. Because of her genetic history, doctors may harvest her eggs, make the miraculous match, evaluate the embryo, make a godlike selection, and implant it. They may intrude upon her cycle and intervene in the lunar one. Or perhaps she will make a different choice. Had we done genetic testing on her before her cells multiplied and she grew into the vibrant being she is, she would not be alive and the world would be a lesser place. With knowledge comes responsibility, sometimes more than we want to bear.

I stand on the shoreline wishing only possibilities of motherhood for my daughter, which happened for me so effortlessly, unconsciously, in a tumble of sweat and flesh one night in a rented cabin by the ocean. As my feet crunch on scallop shells and pebbles, I wonder why I found four or five exoskeletons of horseshoe crabs last week and this evening none. Where do they go as the days grow shorter? On what early morning when I'm sleeping will they know to waddle up on land and spawn, and how many eggs will survive? I have not seen this, nor have I seen the blue blood that has saved millions of anonymous lives.

Limulus polyphemus, how ugly and strange, your name the same as the not so bright Cyclops whom Odysseus outwitted. But then, that giant had one eye with which to navigate, not ten, and that giant drank himself into an unconscious stupor. Odysseus instigated his escape as Polyphemus slept; at another time, Odysseus's men killed the Cattle of the Sun against his instructions and with dire consequences as Odysseus slept. The Greeks advised wakefulness; sleep, it seems, is oblivion, loss of consciousness, loss of control, and for Polyphemus, loss of sight.

Because of the horseshoe crab I wish to have eyes in the soles of my feet or along my ribs or the back of my head since apparently it's not impossible. In deep water the crabs swim upside down, their underbelly eyes picking up light shafts slanting through the water, while eyes on the carapace search for worms and crustaceans below. Imagine! What panoramas! What sights when the wind whips up behind you and flips green leaves to silver while a single leaf flickers down before you, tilting and turning so strangely like the brown butterfly there braving an October wind. You see it all, (and all is irreducible, since the eye cannot block out half a frame) and it wouldn't stop even as the wind dives beneath your conscious mind and whispers unintelligibly of things extinct and things unseen.

Fearful Asymmetry

M Y BEAUTIFUL COUSIN JOAN came to Nassau with us for a week one summer when I was eight. She was in her twenties, and I admired her trim yellow two-piece as she sat on a towel to sunbathe and rubbed baby oil on her arms. She stretched her legs in front of her and crossed her feet as she offered me iced tea from the cooler. It was then I noticed that one of her feet was smaller than the other and curled inward. As I looked away, unsettled, she casually remarked, "Yes, if I get new shoes I have to buy two pairs—size eight and size five." When I asked why, she responded as naturally as if I'd asked about her favorite flavor of ice cream, "Because I have a club foot. I was born with it." Later my mom would explain that as a little child Joan had to wear an ungainly boot to school. As she aged, she walked with a dip to the left and developed back problems from her uneven gait.

When my brother was thirty-five, he developed an acoustic neuroma, a golf-ball-size tumor near his spinal cord that affected the nerves running from his inner ear to his brain and required an eight-hour operation to remove. Prior to surgery he learned that nerves would be cut, that he might lose muscle control in his face, that he might suffer perpetual ringing in that ear or lose hearing altogether. Years later I still remember sitting with him in the living room of our parents' house. He held his head between his hands, staring at the floor, then looked up and said with dread, "One side of my face might be *saggy*." My brother had a strong jawline and deep-set eyes—handsome I realized viscerally as the operation approached. He survived the surgery and retained the structure of his cheek, but as days marched on post-surgery, ringing in his right ear turned to silence. He was a professor of ethnomusicology, proficient at piano and every instrument in the Indonesian gamelan. His world became aurally asymmetrical. When we go for a walk I need to be on his left. At dinners he opts for a seat at one end so that no one sits on the side where his hearing is a void. He doesn't complain or even describe what this situation is like, and if I asked, he'd probably shrug and say, It's a drag. When he and I were little we had record players that played mono; then stereo appeared opening up dimensions of sound, and there was no going back. Mono sliced off a layer of richness, like skimming icing off a cake, or eliminating connotation, or filtering sunlight through shades, or sanding down a paint-rippled surface.

A friend of mine built his house with two identical two-story structures joined by a long empty hallway. In the living room, every painting is paired with another, identically framed, and the sofa is bracketed by identical white tables. Two vases equidistant from the end of the mantel stand pleasantly, accenting the perimeter of the fireplace. All balance, Palladian order and calm. Even the gardens at opposite sides of the lawn add symmetry to the scene, one circular line echoing the other. He has shaped the exuberant hydrangea, daisies, dahlia and peonies to adhere to the plan so that while no plant precisely mirrors another, the desired effect is equilibrium.

In the human form and in that of most animal life, nature favors symmetry with only slight deviations, one breast a bit larger, a freckle here and not there. So I was startled, walking on the beach of Peconic Bay off Long Island, to find a tiny pale crab with one extravagantly outsize claw. Was this *normal?* Would its body grow into the claw? Why was it wielding this thing as it zipped beside a rock where it was miraculously camouflaged and then scuttled into a hole, leaving the claw resting at the entrance? How curious and comical, this lopsided creature, this crab-world version of tennis great Rafael Nadal with his bulging left forearm. In the days following, I often walked along this beach just down the road from my house, and I looked for the crabs. Sometimes there were dozens skirting sideways along the shore and pausing (as if hiding for an instant) in tidepool shallows. Asymmetry abounded! Most bore one weighty claw. I bent down to look closely but they were skittish, starting and stopping, halting by a pebble or sun-bleached shell of similar color to fool me. Probably I would have missed them had I not seen them move. Amputated crab legs and empty pecked-out shells littered the beach—the timid crabs were unnervingly vulnerable carrying their quizzical burdens, looking malformed, aberrant, as if nature had hiccuped and attached the wrong appendage. One darted down a hole but kept his claw at the surface, sand colored and very still. Beneath his beach umbrella (the big claw, in fact, dissipates heat from the body as breezes pass over it) he stayed cool and out of sight.

Only male fiddler crabs bear the massive claw; females have two delicate ones. But with these creatures, unlike Nadal, the smaller claw is the vital one, the one with which they eat, the big one being too cumbersome to reach the mouth. The large claw is not even a great weapon against predators. If it is crushed, another big claw, though more brittle than the first, will grow in its place. Although not a reliable sign of strength, the claw acts as a sales pitch to females, signaling, "Come to my house" (since the big claw won't fit in hers) or a spurious threat to other males, saying, "Stay away." Since it takes energy to grow the big claw and wave it rhythmically, the longer the better, females might find such fitness a winning trait in a mate. Reasons for the morphological development of the claw are puzzling, though, due to its size and weight. If its sole function were to signal to a female like a flag, the claw would be large, long and highly visible, but light enough to sustain without a great loss of energy to the crab. If its purpose were solely to ward off competing males, the claw would be heavy but smaller, shorter, and easier to manipulate. The objectives driving this exaggerated sex characteristic's evolution are in opposition, which is pretty unusual and perhaps a reason for the variety in the structure of the big claw across the hundred or so species of fiddlers.

Blatant asymmetry in animal life is baffling and seemingly random, both in how it happened over eons and why. The little fiddler is not alone in its oddly off-balanced body. Narwhals have two upper teeth, which are ostensibly symmetrical, but in males and some females, one tooth soon grows straight through the upper lip and extends about eight feet in front, looking like a long skinny spear. This is the lopsided unicorn of the sea! A fantastical creature whose tusk seems a comic accident to inspire our imaginations or elicit applause for the vagaries of nature. For a long time scientists thought male narwhals used the tusk just to joust a bit and tell a competitor who's who, not unlike the fiddler crab waving its ungainly claw. It took an expert in dentistry, Martin Nweeia from the Harvard School of Dental Medicine, to figure out a more vital function: millions of nerve cells run from the center of the tusk to its outer surface, allowing the narwhal to gauge water temperature, pressure, and salinity. The tusk can detect variations in the quantity of particles in the water, which helps the whales find food. Drone images show them gliding through the water tracking schools of Arctic cod. With a tap of the tusk a whale stuns a fish and consumes it. Narwhals have superb directional sonar, issuing one thousand clicks per second and detecting echoes bouncing off rocks, possibly through the sensitive tooth: water drops seep through pores at the tip and bubbles travel up the tusk to nerve endings near the head, delivering messages to the brain.

With their sensitive spears, the males enjoy "tusking," or rubbing tusks, which definitely sounds erotic and produces sensations we can only dream about in another world—human teeth rubbing not being so desirable. With their sensitive spears the narwhals also fight, suffering clashes that put at risk the whale's ability to feed itself, to know where it is. The potential for pain and pleasure, life and death, are inextricably connected in a seemingly aberrant tooth.

One evening, after watching a sunset flood the wetlands behind my house, leaving telltale strips of ocher and rose along the horizon, I sat on my bed staring at two photographs of the ocean. They were mounted in identical frames on the opposite wall. A nice pairing, I had thought. But one image of a cresting wave flinging spindrift and poised to crash did not amplify the other, a fist of water pounding the shore and spitting up froth. Bleached wood frames hung side by side undercut the violence, lending to the two scenes an aura of control.

If symmetry is generally comforting and empowering, why did William Blake find it fearful? In his famous poem about a tiger, or about God making a tiger, he addresses the tiger, wondering, "What immortal hand or eye, / could frame thy fearful symmetry?" Crafting a symmetrical framework for the poem, he ends it by asking once again, with a minor variation, "What immortal hand or eye, / dare frame thy fearful symmetry?" Blake knew as many writers do that there is comfort in return—the return of the hero, the return to birthplace, a reiteration at the end of some idea posed at the start. But he disrupted that comfort by leaving the questions unanswered as well as by assuming we're ready to understand why symmetry might be fearful. According to the OED, Blake's use of the word *symmetry* meant not an exact mirroring of elements, but "form," with an implication of that form being "well proportioned." Well-proportioned arguably suggests balance since by the mid 1800s *symmetry* came to mean a correspondence of parts on both sides of an axis, while today Merriam-Webster's first definition is "balanced proportions." The question remains, why did Blake write that this form (with its potential to be aesthetically pleasing) inspires fear?

We might assume *fearful* to mean *fearsome* and attribute it to the tiger, as if it's a given that we'd all be afraid if we met one; alternatively, if *fearful* means *afraid*, it is the tiger that is scared, not us. In fact, *fearful* clearly modifies *symmetry*, and the statement is counterintuitive. We stare with morbid fascination at a clubfoot, fear deformity and disfigurement. Blake gives symmetry enormous power if he wonders how any god *could* frame it or *dare* frame it, meaning embrace or control it. And if no god could, then in Blake's theological paradigm that symmetry might exist outside of God's will or even prior to God's creation since the question is how to frame it—just as we have a painting first and then wonder how to mount it. What frame would do it justice? If each tiger's markings are individual, then so should be the frame, making identical frames or symmetrical positioning of frames an injustice. Or maybe symmetry is synonymous with perfection, a leap, but a possibility, suggesting that we fear it because we can't attain it ourselves. But why would a tiger be perfect? And why more so than a lamb (which is also mentioned in the poem)? The disjunct in power between tiger and lamb is so vast that the incredulous poet wonders if the same god could have made both, leaving the reader with a world off kilter, out of balance, asymmetrical not only in its inhabitants but in the powers that created it.

One day follows another on Peconic Bay—little fiddler crabs scuttle along the shore and sunlight emblazons the water—and the water that day was sapphire because the sky was clear. Endangered plovers were quiet in the sere grass, no dogs racing through to disturb them. Usually the beach was deserted though one might find old Mrs. Crenshaw with her three scruffy mutts or a guy tossing a tennis ball for his lab to retrieve from the bay. Now a handful of people were there, all standing rather than strolling or picnicking, and all eyes were on the horizon. Why? A stocky woman in a black bathing suit cut low between her breasts was shielding her eyes and gazing out. A boy, about twelve, sat sun-pink on the shore, breathing deeply from a long swim. What was out there?

Someone told me it was another boy, this one still on his paddle board. He'd been trying to come in for so long and had made no progress. Someone told me she called 911 and the police appeared. I strained my eyes—had no one told me I would have missed him—and there just a few feet from the horizon was a pale figure, alone, one paddle at his side. The cop called the Coast Guard and we all waited.

Wind that had felt pleasant a few moments before as it dried the sweat on my stomach and back suddenly seemed brisk—a determined wind, sweeping offshore. The boy would not make it back, that seemed obvious. Why did the stocky woman, the *mother*, just stand there, gazing? I would start swimming. Maybe she knew she wouldn't make it. We waited. The greenheads were fierce, stinging my thighs before I saw them, so I walked down the beach to keep moving. Finally a green boat no bigger than a toy pulled up, and the figure vanished from sight. They must have him, we sighed, and watched the boat advance, at last delivering the boy to the shallow water just off shore, a light-haired, round-faced, half-naked angel. He was about ten, smaller than the other one. His face betrayed nothing. He had no life jacket.

A week later, on two successive days, two women drowned off nearby Long Island beaches. The waves on the south shore had been choppy but not overly so, nothing to cause alarm. The water was finally warming up—it was so inviting on a sultry day. But under the splashy waves and froth, a riptide seized these women close to shore and they could not clamber out; even where their feet could touch bottom, they could not get out. I was there on the second day, unwitting. Only as I left the beach and five cop cars screamed down the highway did I realize something had happened. I had sat there on the beach while the waves rolled in, took her, rolled in again, even as I bit into my chicken sandwich.

⹀

I don't wonder as philosophers have done for centuries how God could have made the miracle that is the world and included so much suffering. I wonder why we presume that the world would be designed to avoid it, our human pain, why we assume our dichotomies of beauty and violence would align with those of a divine creator. Birth and destruction, joy and horror, are intertwined asymmetrically. I don't believe there is equilibrium in the universe; it's off balance like my beautiful cousin, and we hear only a portion of what it has to say, like my brother, even though our legs carry us up mountains and our ears are so nicely aligned.

Asymmetrical patterns surround me—not claws or spears—intangible ones. Hummingbirds alight, stop, swerve, buzz, dip in, spin out of sight, leave a zigzag trail of absence. Wind turns lake water black, shifts, leaving boats stranded, and lies down for the night. I wake to a red-winged blackbird's insistent trill. Does another answer *chit, chit, check check* from a farther point? A catbird mews, and a swallow chippers and whirs. A friend buys a house, another loses his to fire. The reasons elude me.

As I walk down the beach, two sandpipers run headlong, lift off in sync, trace the same arc five feet over the water, turning their white breasts to the sun, and land at precisely the same moment. Graceful, occupied, they move a safe distance from me in seemingly effortless partnership, leaving me to wonder if that is all we want, after all.

Dead Man's Fingers

I N THE SHALLOWS OF Peconic Bay, seaweed clouded the stones just under the water's surface, seething and tilting in the slight waves. Ella, at fourteen months, halted, pointed, and wouldn't even walk around what seemed a solid mass. "Seaweed," I said. "It's okay." Wading through tidepools, she pointed at clumps of olive-green rockweed with its puffy air bladders or sea lettuce glinting bright green. I picked them up, slimy and cool in my palm. "You can touch it."

For weeks she refused all species, diverse as they were, eyeing me obliquely. We always found spongy green branches of seaweed, anchored to small rocks and sealed under slipper shells. "It looks like pasta!" I remarked, hoping a familiar association would help.

"Patta," she repeated.

"Ella, you can touch it!"

"No," she blurted, weaving past me and past rubbery strands splayed in the sand. I told her the seaweed needed water to live and flung it back.

A few days later, sighting stringy seaweed along the shore, she exclaimed, "Patta!" I wondered if I had confused her by introducing metaphor at that point in her life, until she told me back at home that water pouring from the bathtub spout was "like a waterfall."

The common names for this spaghetti-like seaweed, *Codium fragile*, are in fact pure metaphor: dead man's fingers, green fleece, staghorn weed, and oyster thief since, with buoyant strands filled with oxygen, it can lift oysters from their beds. It's also called Sputnik weed, after the satellite the Russians launched in 1957. Someone watching on Orient Point at the tip of Long Island that October found some green algae never seen there before—or anywhere in the country. It turned out to be *Codium fragile*, which had made its way around the globe from the Northwest Pacific and up the southwest coast of Africa to Europe and across the Atlantic to us, thereby assuming the rank of invasive. Meanwhile, the orbiting satellite emitting eerie blips from a night sky provoked an outcry from Americans, terrified about their privacy and safety. For Ella and the seaweed, I didn't know if it was movement where she didn't expect it or her inability to see what lay beneath dark masses undulating in the shallows that caused alarm.

Not wanting her to fear anything in the natural world, I persisted. When I held an emerald bit of sea lettuce in my palm one morning, she finally touched it with the tip of her index finger, then went to look for minnows. She poked her finger down hermit crab holes, lay down in tidepools, and struggled to pick up ungainly rocks coated in algae.

One day weeks later she went into the water and scooped up a handful of sea lettuce. Another day she kicked aside a congregation of rockweed. Then she found a sprig of dead man's fingers in the sand, picked it up and toddled to the water where she tried to immerse it, and commanded, "Live."

During the spring of 2020 fear sprang from unexpected places, both erratic and irrational: fear of a woman next to me in the bagel store whose breath might just have nestled in my nose, fear of my brother who got off a plane and came to dinner, fear of the bill in my mailbox, touched by whose hands? It was guerilla warfare. Viruses could be anywhere, probably were everywhere. In March my daughter and son-in-law had left Brooklyn with Ella and had come to live with me on Long Island. As a toddler, Ella absorbed the effects of tension and estrangement, not yet the cause. When a figure came down a deserted stretch of beach, she'd climb into my lap. "Man." "Yes," I confirmed, "man." After he passed, she would ask, "Where'd he go?" And I would say, "I don't know," and she'd wait till he was nearly out of sight before leaving my arms.

In early fall I drove to a small town in Westchester to visit a friend. Together we walked along the Hudson River about ten miles north of Manhattan. On a strip of muddy shoreline between the river and a parking lot lay scores of dead menhaden, eyes blank as nail heads, their bodies whole and wasting, askew, like scrap metal. A few seagulls sat on pilings in the river. None came near the fish.

I learned that dead fish lay strewn on shores from South Jersey to Rhode Island. Heated water, lack of rain, sewage, and algal blooms suffocate fish. Suddenly, they can't breathe, an unnerving image, evoking wildfire smoke, my grandfather's emphysema. New York's Department of Conservation commented, "This type of fish die-off incident occurs naturally this time of year when water temperatures warm and generally have little impact on region-wide fish population numbers."

Will we say of 2020 that 316,844 COVID deaths had little impact on U.S. population numbers? It would be true.

Scientists found vibrio bacteria in the flesh of the fish in that die off. A new cause of death, or an additional one. And Delta arrived.

===

With warming water in ponds and lakes, algal blooms swirl in blue-green, yellow, rust, and brown, thicken, and scum over surfaces with cartoon green. When nitrogen and phosphorus leach from soils into water and heat up, cyanobacteria populations boom. Witless dogs and livestock step in for a drink and either die within hours or are left with a list of hideous symptoms—bloody diarrhea, mental derangement, photosensitivity, convulsions, breathing issues—on par with potential side effects of Lunesta, advertised on prime-time TV. The wealthy are not immune. A dog died in East Hampton's Georgica Pond; a boxer named Bella died from the waters of the C-51 Canal in West Palm Beach. Cyanobacteria flourishes and toxifies bays, estuaries, and lakes on Long Island, turning scenic water that once spiked real estate values on the East End into fetid ponds to be feared.

Don't go near it, Ella, don't even touch the shallows with your finger, I will have to say. And that may trip a wire, light a fuse, ignite fears that billow and bloom as she questions, why, why, why. Why pesticides, why is it green, why is it orange, what do they do, what is sewage, why do fish die, will I die?

===

January 5, 2021. I lie in bed for the fifth day and chug glasses of water and eat more than I want and believe I can will away the fever. I don't think I won't get well, but words skitter through my mind—*fever spikes, salt smells, lung collapse*—while overhead or in the walls, I can't tell where, I hear scurrying and thumping, creatures—how many?— chasing each other, dancing, playing tag—*thump, thump, scriff, shoosh*—bigger than mice but what, their playing a tease as I lie there, nose burning. This virus may be immortal, longer lived than the trees that whirl outside my window, telling me winds gust at fifteen miles an hour or more. I have time to watch. I have COVID. As I lie sweating through my T-shirt and voices on TV lose coherency, knobby virus cells effuse around the globe like an oil spill and suffocate like the tepid water of the Hudson... Russian hackers infiltrate the country's backbone... the electric grid could blow... lights go out, medical services freeze...banks shut down.

Badat! Badat! rang Ella's voice on the beach in early spring, before all this. Dimly, in half dreams, I see her pointing at the sky. What did she mean, what was she saying? *What's that? What's that?* Airplane, Ella, airplane.

She came to use the two words interchangeably. An *airplane* was a *badat, badat,* an *airplane*—coalescing question and answer, uncertainty and affirmation—as if we would know, as if we too could recall why we learned what we know and how a thing became a metonym for the impetus to know. Such a fusion might salve the terror of things unknown, which orbit our minds in a fever or at 3 a.m.

Before ten days have passed, I step out my back door. Winter ice shines on the road and over snow. Peconic Bay lies absolutely still—without a breath, or holding its breath. The shallows are clear of seaweed, most of the gulls gone. Far out the black silhouette of a loon rests like a question mark on the surface, vanishing from time to time and popping up somewhere else. Underwater it rockets after sunfish and croakers, swallowing a fish whole and headfirst. On the sand I find forgotten masks, not condoms, twisted among dried clumps of rockweed.

Was it the random movement of a brown mass around her toes that had scared Ella, or the loss of the sight of her own feet? She had stared at seaweed every day until perhaps the pebbles under her soles became an act of faith. Without science or metaphor, she connected fibers undulating eel-like in the shallows to green branches lying inert on the beach. A count of 316,844 deaths reveals nothing about the life of one victim, and I don't know if knowing one victim would connect me more tangibly to the rest. But stepping without seeing what lies below or ahead is what we do because we have to.

The Midnight Zone

1.

Hearty gulls drop whelks on a stony beach near my house, again and again, till the shell cracks and they access the snail inside. Skeletal whorls remain, or deceptively whole shells tucked in the sand, concealing a slit at the keel or ragged edges to the outer lip. Striated, barnacle covered, gray curves pecked by gulls, they are never unbroken. A spiral with a sheen of peach and coral lies among the rest and I wonder how it was so ravaged.

A whelk roves the bay floor looking for oysters, clams, and crustaceans. It detects water streaming from the feeding tube of a clam and descends, bearing down on the clam with a hard foot that protects the snail inside like a tough oak door. The whelk pries open the bivalve with the outer lip of its shell, even chipping its own edges in hot pursuit. It manages a crack through which its proboscis slips in and begins to feed. A whelk will cannibalize other whelks: it secretes a fluid that softens the calcium carbonate of the shell and then bores a hole with its long-toothed radula, which reaches into the body of its prey and retracts it, bit by bit. This is the less visible history of broken whelks.

2.

In 2007 Inuit hunters killed a bowhead whale off the coast of Alaska using a bomb lance that pierced the whale's blubber and exploded inside. When they carved up the whale, a biologist found a similar weapon embedded in its neck dating from the 1880s. The harpoon fragment is on display in the Inupiat Heritage Center in Barrow. Biologists verified the age of the whale by studying its eyes, which grow cloudy with age, as ours often do.

From November to April when Arctic waters are darkest, the bowheads sing continuously. Some warble in a rising arc, others moan or groan or rumble like a deep long-winded Tibetan horn. Of 184 recorded songs, no two are precisely the same. The whales improvise and invent, creating phrases that shift in volume and pitch. Palpable shapes of sound roil through the water, call it the sight of sound. Sixty-foot whales with giant baleens navigate by the moans of other bowheads or alien calls of fin whales and humpbacks that are moving north. I like to think a female relays to her young the hazards of a harpoon. More likely, she is gauging the strength of her Pavarotti of the deep by checking out the duration of his song, its intricacy and coloratura. But this seems reductive, or at least not the whole story.

When ice melts, wind can create waves on the water; when waves break they make bubbles that pop. Bubbles can change a soundscape. So do cruise ships. The bowheads are listening. They continue to sing for months and months and compose new songs, perhaps painting aural vistas of rising air and heaving ice, obscured by diesel electric generators and pummeled by propellors.

Although no one knows if Homer existed, all iterations of him are as blind; poetry has no need of the visual—it's music. It can tell an epic story.

3.

Jellyfish tentacles read changes in light and temperature and detect the presence of others. They have no brains or eyes or ears, and their skin is so thin that oxygen diffuses within, without any need for lungs. Wispy and filmy, jellyfish pulse and billow throughout the world's oceans where they began about 500 million years ago. They thrive in warm, overfished, polluted water. In 2013 a bloom of moon jellies clogged the pipes of the massive Oskarshamn nuclear reactor in Sweden and forced it to shut down for two days.

Jellyfish begin as free-floating larva drifting through the ocean until they land on a firm surface. They anchor like an anemone and grow into a polyp, a small stalk with tiny buds that eventually break off, feed, and transform into a medusa stage—a globular form with tentacles so named because of mythological depictions of Medusa with writhing, snake-like hair. Although fragile and soft, they have muscle power to navigate and stinging cells with venom in their tentacles to stun prey.

The species *Turritopsis dohnrii*, about the size of a small fingernail, has assumed the status of the immortal jellyfish. If stressed, damaged or starving, a full-grown medusa can shrink into itself and reassume a polyp form where it awaits favorable conditions in which to bud. The new medusas, while genetically identical to each other, differ from the original medusa. The jellyfish can transdifferentiate, or go back and forth, many times. *T.dohnrii* was first found in the Mediterranean but is now global, partly due to its ability to revert to polyps in the ballast water of ships without much food and regenerate on arrival.

Despite their power in numbers to wreak havoc and claims of immortality (naturally of interest to humans) and ability to regenerate (auspicious for methods of healing), jellyfish are 95% water. If washed up on shore, a quivering, shiny orb of magenta and rose petals may evaporate in less than an hour and leave nothing, or nothing a passerby would notice.

4.

For decades scientists debated which of the five ancient animal lineages—placozoans, cnidarians, bilaterians, sponges, and ctenophores —is the oldest. They narrowed the choice to the last two. But even complex genome sequencing couldn't provide a convincing answer to which one branched off from the first multi-cellular life existing hundreds of millions of years ago. Fossils of soft-bodied animals living long before the dinosaurs are extremely rare.

A sponge has no muscles or nervous system and after its free-wheeling larva stage, it sits on the ocean floor and basically waits for food to pass through its filters. A comb jelly (ctenophore) is a diaphanous creature with an oval or bell shape and an elegant arrangement of eight strands whose cilia propel it through the water and diffuse light to shimmering effect. Ctenophores are avid predators that catch prey with sticky cells. They have nerve nets, and some are bioluminescent. In our dim imagining of origins, the primitive sponge seems likely, the comb jelly more evocative. Autonomy! Light! Agility!

In 2021 a graduate student at UC Santa Cruz, Darrin Schultz, and his co-advisors established the chromosome structure of a ctenophore. Experiments followed at UC Berkeley in which scientists discovered that sponges shared an arrangement of genes with other animals whereas the comb jelly's genetic codes were different. Its gene-chromosome combinations resembled those in one-celled organisms, which Schultz and the Berkeley team maintain, means that the comb jelly branched off before the sponge's evolutionary rearrangement occurred. Exclamatory announcements followed in 2023 about the oldest living creature on earth. And since ctenophores date to 700 million years ago, sponges to about 600, we animals may be even older than we had thought. Naturally, there are skeptics.

As I look at the ocean, I imagine traces of my DNA held fast in tough coastal ctenophores that tumble in the waves, in others that glide through open water—species too fragile to gather—and still others wandering and hunting through the midnight zone. So distant and amorphous are they that we've given them tangible, folksy names like sea walnut and sea gooseberry. Or sexy ones like Venus's girdle. The connections are incongruous, out of sync. Meanwhile, I search my cluttered garage for a sponge, the natural kind, with which to wash my car.

5.

Languidly, Greenland sharks glide through chilly waters of the Arctic and North Atlantic, from Baffin Island to the Barents Sea. They are not picky, eating other sharks, seals, eels, whale carcasses, seabirds, and an occasional reindeer that falls through the ice. Marine biologists discovered that certain proteins in the Greenland shark do not renew, so tissue exists that was formed when the shark was a pup. Taking isotopes in the nuclei of the eye from twenty-eight sharks caught inadvertently in fishermen's nets, a team from Copenhagen used radiocarbon dating and found the sharks to be 275 to 500 years old, the oldest vertebrates on earth. What time map propels this creature that dives slo-mo to the midnight zone 6,000 feet deep under pulverizing pressure and doesn't even feel the urge to procreate until the age of 150? There is hunger and satiety—like whole notes and rests—a long, elastic, homogeneous existence.

Even with centuries on her hands, a female doesn't tend to her young. Off the pups go, alone, to mosey through the cold and dark. Lacking interaction and rites of passage, this state seems a kind of unbounded solitary confinement in open ocean. Lacking interaction, my nearly centenarian mother nods and sleeps and half sleeps while wheelchairs circle around her and aides carry trays and visitors come and go. That is the

unplumbed loneliness of old age because who in that state can articulate
where the mind travels through an afternoon, how the mind conceives
of an afternoon if it does at all. The just-born baby shark glides off alone
to eat; my mother's near-death mind swims through images of that day
or a day some time ago. Out there, down there slides a shark that might
share a birthday with Shakespeare or have seen the *Titanic* go down, its
healthy heart beating just once every twelve seconds.

These fish have no white blood cells to attack viruses and bacteria.
The older ones have various combinations of gene mutations that fortify
the immune system. Scientists want to develop drugs that mimic these
genes, or modify the human genome to do so, giving us a greater shot at
longevity through an intimate connection to this shark.

6.

In 1964 Donald Currey, a geography grad student from the University of
North Carolina cut down the world's oldest independent living organism with a chainsaw. Or possibly the oldest.

Scientists knew that bristlecone pines growing in Nevada's Snake
Range at the timberline 10,700 feet above sea level had been there for
millennia. Currey was researching climate change with a focus on the
"Little Ice Age," beginning in the 15th century. He wandered around looking for a specimen that could reveal data from that period, found his tree,
and named it WPN-114. Easily anthropomorphized (and later mythol-
ogized) the chunky old bristlecone with a twisted trunk and shock of
ragged leaves clung to the glacial moraine with huge gnarled roots. Only
a 19-inch swath of bark remained, the rest eaten away by ice storms. It
was seventeen feet tall, mostly dead, but with one live eleven-foot limb
due to a phenomenon called "sectored architecture." If some roots die,
only the part of the tree just above those roots will die. Other roots can

continue to feed other sections of the tree. This construction, along with the twisting branches that sync with wind from any direction, allows the tree to live far longer than trees with a single support system. To count the rings and determine age, Currey inserted a bore into the center of the trunk and tried to remove a narrow strip of core (cringingly reminiscent of CVS testing), but it got stuck or broke or was too short—accounts of the story vary. He turned to the Forest Service, which granted him permission and provided a chainsaw team to verify the antiquity of his find. Currey told *Terrain.org*, "A horizontal slab from the interval 18-30 inches above the ground and a smaller piece including the pith 76 inches above the ground were cut from the tree, and a smoothly finished 2-piece transverse section was prepared." His euphemistic language, which omits any active agent (himself), cursorily outlines the demise of WPN-114, later said to be about 5,100 years old. For decades, the act was controversial, more the stuff of horror movies than a study of life.

Yes, Currey could have figured out the age of the tree without slaughtering it. Yes, he could have focused his research on his topic rather than the age of an individual tree, a sensation that distracted both him and the public. As heinous as the act was (one veteran Forest Service sawyer refused to participate), the aftermath was as tangled as that old tree's branches. Myths arose that touching the wood of the bristlecone would lead to early death. The tree was renamed Prometheus, a martyr to human desire for knowledge, a visceral image of one gashed and suffering, alone.

Humans' thoughtless ravaging is an old story by now. The emergent story, the urgent one, lies in comparing this tree's rings with those of other ancients to place each in a chronology of warming years, of movement of the species up and down mountains, of droughts and floods and volcanoes, health and illness, of *connectedness* to environment at any given moment. The pith and rings reveal not only a numerical record, but also less visible histories. Anyway, what are the odds that Currey found the oldest tree in that grove?

Before this incident, about fifty people had seen Prometheus. A 4,700-year-old bristlecone stands on eastern California's White Mountains, but its precise location is guarded to protect it from us.

Possession

CRACKED THE WINDOW and hot air whooshed at my face. Miami suburbs slipped by and dissolved as the horizon opened up and my rented Honda skimmed along like a bubble on a river. The only voice was from my Google Maps app, warning me matter-of-factly about when to exit and which lane to be in. Route 41 stretched ahead, straight and fluid, cars shining in the sun and moving with unfamiliar calm. After just a few days, I was leaving Miami behind, its persistent line of hotels that loomed over the beach like too many wedding cakes, and balconies that ballooned toward the sea, its beaches throbbing at spring break, young women sporting thongs tucked between buttocks looking like unpicked watermelons.

I was going to see alligators. Along with a line of cars and tour buses, I turned into the Shark Valley Everglades Visitor Center where I rented a clunky bike with large handlebars, back-pedal brakes, and a handy basket in the front like those I had as a kid. A narrow roadway loops out about eight miles to a lookout tower then heads back through open vistas of sawgrass and wild tamarind. With a bottle of water and my mother's opera glasses, I chugged along, sweating through my sun-proof shirt. The path followed a sluggish waterway about five feet wide overhung with red mangrove, wax myrtle, and fetterbush. Sporadic groups of people lingered, oohing over a snapping turtle, a blue heron. Then I heard *alligator* and stopped. Camouflaged in the shadows, it lay listless and still, half its body on a rock as if clambering all the way out had been too much. I began spotting others, also motionless despite the proximity of human flesh. Alligators can sprint up to thirty miles an hour, but why bother? Around them lay a rich menu of fish, crabs, turtles, birds, and berries readily at hand. Only one slithered just under water, slow and deliberate as the octogenarians with sun-leathered skin in the 87-degree pool at my hotel.

Life is good at the top of the food chain. I thought of nearby Palm Beach where I'd stayed with a friend for a few days, just north of Trump's heavily guarded enclave. Bentleys cruised at twenty miles per hour and bodies languished by pools. Houses looked eerily empty. Only the landscape crews bustled, carting coconut tree debris from pink mansions, clearing carpets of shaved grass, and trimming squared-off hedges that kept invisible and hushed whatever predatorial behavior it took to get there.

Now the hot breeze felt good on my face, the sky broad and empty. I pedaled on, glimpsing wood storks and egrets amongst the thick mangrove stalks. I stopped to take pictures of alligators, zooming in for maximum shock value. Friends texted back: "omg—don't get eaten, *where are you?*"

Any answer seemed to rely on the alligators, their blank eyes, their armored bodies, their large clawed feet that look weirdly like hands. Watching a six-foot reptile swerve through a swamp conjures a shadowy time of picture-book dinosaurs and volcanoes, the absence of *us*. It is the reptile right before you and the plastic alligator you played with as a kid and battled with your friends' lions or dragons. A titillating object of repugnance, its menacing upturned mouth, a crafty "smile." It is flash-card images of itself in your mind, darting from the water and sinking its eighty teeth into your leg, into the feather-light body of the spoonbill watching beady eyed from the shore. It is a child's objectification of fear, of first awareness that animals eat animals. It is a belt, a wallet. Meat for humans and pets. A familiar form coupled with an unfamiliar place and dislocation in time, which is a recipe for the uncanny, which demands we pay attention. Just thirty miles away, traffic seethed on the highways reaching in and out of Miami.

When I heard about the pythons some weeks later, my construct of alligatorhood shifted, like a tachymeter bezel on a watch. In the early '80s the exotic animal market boomed in southern Florida. People bought iguanas, marmosets, peafowl, swamp eels, giant capybaras, and Burmese pythons, often with no idea how to care for them. More than a few were surprised that their python could stretch the length of their twenty-foot living room. What to do? Let the pet loose in the Everglades where it would surely thrive with ample food within its grasp. And many did, just how many, no one yet knew or cared. The trend for possession of the exotic blindly continued. Some avid merchants in the exotic reptile business decided to *breed* pythons in a rented, flimsy greenhouse in the city of Homestead, twenty-seven miles east of the Everglades. There, baby pythons, a few inches long, curled in plastic cups, alongside scorpions and tarantulas. There they were, cracking out of eggs, wound in their cages, virtually spoon fed, when Hurricane Andrew struck in 1992, its 175 mph wind taking down 25,000 houses across the state and catapulting the baby pythons westward. While most died, some apparently survived to start a comfortable life in the swamp where they grew, ate, and bred—feeding on raccoons and rabbits and muskrats and bobcats, and laid 50-100 eggs per mature female each year with nothing to stop them except the occasional very large alligator. No one knew much about the pythons until war broke out in 2003 near the main visitor center. People strolling on the Anhinga Trail suddenly witnessed a python coil itself around an alligator, which retaliated in a frenzy, laying its vise-like jaws into the snake. The two wrangled for twenty-four hours before the wounded python gave up and slid away. More often, it is the alligators that fall prey, and their size is not a deterrent. With their elastic jaws, pythons can swallow a deer whole after a strike to the neck and a suffo-cating squeeze.

By 2012 the USGS reported that in a fifteen-year span pythons in the southernmost part of the Everglades had decimated the raccoons, opossums, and bobcats (their populations declining by 99.3, 98.9, and 87.5, respectively). Rabbits and foxes were gone. Tens of thousands of pythons, possibly many more, occupy this territory. Finding them to count or to kill is hard given the lack of roads and inaccessibility of the swamp, though a counteroffensive is on, championed by volunteers, amateurs, and professionals trying multiple tactics from shooting (which is illegal) to laying wire fences to training dogs to pick up a scent. The Florida Fish and Wildlife Conservation Commission hosts a ten-day Python Challenge every August when most hatchlings emerge, and the state pays about a hundred contractors to trap them year round. The situation is so desper-ate, scientists are working on genetic ways to kill off females or cause all offspring to be male. The snakes are advancing north into Georgia.

The sanguine alligators I saw might not yet know that their kingdom has been toppled, that they no longer occupy an inviolable seat at the pinnacle of the food pyramid. But humans do, I do, which yanks them from their Edenic, primordial existence and tosses them into contemporary global conflict. The alligator is involved in the messy web of snakehood via a Burmese snake, the other, a satanic trespasser and bearer of knowledge. It is caught in a web of human carelessness and vanished rabbits and foxes where the python survives, eradicating pasts and unfolding futures based on perpetually changing recipes.

In Florida the pythons keep company with a legacy of invaders. Cuban tree frogs eat smaller native tree frogs. Lionfish from the Indo-Pacific devour reef fish that eat algae, allowing algae to overgrow and suffocate coral. Spanish conquistadors introduced smallpox and war with native peoples; the feral pigs they brought plow up forest floors. Redbay ambrosia beetles, recently arrived from Asia, snuggle under the bark of native swamp bay trees where they grow a crop of fungus to eat—a fungus that kills the swamp bay, vital for the lives of Palamedes swallowtail and spicebush swallowtail butterflies. Power structures and relationships are ceaselessly on tilt.

Species I've never heard of go extinct every week. The millions of small parts of the whole are not greater in some objectively valuative system than the whole, just as what an individual does in any moment usually doesn't matter. But it might matter later. Especially to those in the quirky web of contexts linked to that individual. It matters, paradoxically, if you are Putin for whom the whole is greater than the parts and you believe you have the power to affect it, one morally indifferent act following another, as you seize and constrict like the pythons: pregnant women, old men, young men, prisoners of war, children in the interests of possession.

Predatorial aggression is the norm, collateral damage "unavoidable." We dropped 20 million gallons of herbicides on the forests of Vietnam, Cambodia, and Laos so that humans couldn't take shelter. Nor could paralyzed monkeys, scorched tigers, gaurs, pangolins, mangrove seedlings, slow loris with stunned yellow eyes, blue ants, and moon moths… the destruction branched from a singular decision to a pilot to as-yet unborn children who would appear without eyes, feet, fingers, and harbor cells that would grow into cancers to bloom at a later time.

I shudder as I picture pythons cruising the swamps of the Everglades intent on their kills. Inevitably, these images fuse with mythologies of evil in which a serpent plays trickster, where one act of temptation triggers our perpetual suffering beneath a glaring knowledge of impending death. But we took a bite of that apple, after all. We chose, and choose to blame the serpent as we lie coiled in an inability to escape or fathom the violence of our predicament.

In cultures of Indigenous North America and Ancient Greece, in the teachings of Buddhism and Hinduism, snakes play potent symbolic roles, more nuanced than that of the denizen of Eden. Since snakes shed their skin and grow a new one, they may connote both death and renewal, giving grounds for hope. Along with outgrown skin, snakes cast off unwitting parasites; thus, molting becomes a tactic of survival.

In an ironic twist, Florida's pythons are proving a source of hope for war veterans suffering from PTSD, men and women accustomed to working in the wild and themselves practicing crucial tactics of survival. Tom Rahill, founder of a nonprofit called Swamp Apes, leaves his IT day job and heads to the swamps at night, the best time to sight the oily gleam of the snakes. He brings with him men having trouble finding purpose in their lives after returning from war, many from Iraq. Out there, the search is specific, the challenge of battling tough terrain, thickets of mosquitoes, and intense heat is absorbing, the camaraderie a comfort, the skill needed to land a python a source of pride. Some hook a snake and jump on it; others use only their gloved hands, grabbing its back and calming it by rubbing its belly. They bag the python and euthanize it painlessly within twenty-four hours. As one former Marine told the *Miami New Times*, "Riding around, even if we don't see anything, it forces you to be patient.

That helps for anybody who has anger issues or is quick to react. You can't control nature, so just shut the fuck up and keep riding around.... And when you actually catch one, the adrenaline rush is almost like being back on the battlefield."

Despite formidable efforts to restore a natural balance, pythons stand to rule the Everglades for the foreseeable future. Like us, their numbers are exploding. But having devoured so many populations, will they finally go hungry and turn on themselves, gobbling their own tails like the ouroboros? That image strikes me not as a symbol of procreation (phallic tail, wide open mouth/womb) nor of eternity (a continuous circle), but of destruction and desperation. Stressed, disoriented, hungry, overheated, constricted in space, snakes sometimes do devour themselves.

———

Behind my house on Long Island, ten pitch pines design the sky. When my three-year-old granddaughter wakes up here on a March morning, she looks out the window and announces, "Look at the evergreens there!" Just a mile south, 20,000 pines have been taken down because of a beetle that journeyed from the Southeast to enjoy warming climes in the mid-Atlantic states and New England. Cool winter temperatures work to the beetles' advantage since larvae stay on hold till spring and then burst out en masse rather than trickling out throughout the year. Summoning one another with pheromones, thousands of beetles join forces to tunnel under the bark of an ancient pine, digging trenches that cut off its nutrients and overpower oozing resin, which is the tree's only line of defense. It's a synchronized attack and lethal. When I drive through that recently cleared area, the light seems brazen, the shingled houses angular, stark, and vulnerable, as if trees and houses once shared an unspoken symbiosis. When my trees go, I may see more starlight and just get used to that. My granddaughter may forget they were ever there. For now, we talk about the evergreens, eliciting their lives, filling the air between us with what she knows of trees and tree things, just as she scribbles within the lines of a coloring book, spilling erratically outside and onto a page that follows.

Dependence by Design

O N A RECENT BREAK from New York winters, I flew to Florida hoping to swim in turquoise water touted on ads in the subway. As I strolled on beaches, dodging plastic bottles and bags, I saw hundreds of nautical-blue bulbs, some with long streamers running down the sand. "Watch out, don't touch that," warned a lifeguard, raising his substantial biceps to show me a red mark on his inner arm. "I got stung this morning."

"Too many jellyfish for swimming?" I asked.

"They're not jellyfish, they're Portuguese men-of-war."

So I learned that the dimpled half-moon at my feet was not, despite its tangled tentacles, a jellyfish but a siphonophore. Does the difference matter? Jellyfish, which belong to the subphylum Medusazoa, usually have a gelatinous bell-shaped body, which they can contract and expand to propel themselves through the water, tentacles streaming behind. Some might sting but it's often mild and not lethal to humans. As a siphonophore, the Portuguese man-of-war has both medusa and polyp stages of development. Polyps differ from medusas in that they are tube shaped and often immobile, like coral and sea anemones, their mouths and tentacles at the top looking up through the water. The man-of-war, *Physalia physalis*, has its own mix of these forms and is not rooted to the ocean floor. Even more distinctive, it is not a single entity like a jellyfish, but a colony of four specialized parts. It floats because of the blue sac I saw, still puffy on the beach, like a folded dumpling. Filled with carbon monoxide and air, this polyp, called a pneumatophore, bobs about at sea, deflating when it needs to submerge, but otherwise relies on wind and currents to move. Beneath the bulb hang polyps (or zooids) that form a dazzling, curly array of purple, pink and magenta tentacles streaming thirty to a hundred feet through the water, looking delicate and deceptively benign. Called dactylozooids, they can shoot out barbed threads with toxins that paralyze fish and crustaceans. The sting can be intensely painful, even lethal to humans, and these defensive zooids strewn across a beach like strands of hair can still perform if you touch them. (The blanket octopus, which somehow developed an immunity to the venom, will rip off a man-of-war dactylozooid for its own use, warding off predators and attacking prey.) Once the dactylozooids sting a fish, the man-of-war mouths (gastrozooids) go into action, with as many as fifty organisms opening, covering, and gobbling prey that the venomous tentacles serve up. They digest the fish both extracellularly and intracellularly, finally offering the digested material to other members of the colony in a main gastric cavity, the dining hall—the creature lacks a stomach.

When it's time to reproduce, Portuguese men-of-war of both sexes gather around for the action; specialized polyps on each, called gonozooids, disperse either eggs or sperm into a column of water, a phenomenon called broadcast spawning. There, floating about, sperm fertilize eggs and begin to grow as larva into a larger organism. At some point in its development, individuated parts sharing the same DNA bud asexually to form the four necessary components of this siphonophore: air sac, zooids bearing toxic venom, zooids capable of digesting, and zooids responsible for reproducing.

In ecological terms, the colony acts as one individual, but embryonically, as four. Casey Dunn in *Current Biology* puts it this way: "Being a siphonophore is as if you were to bud thousands of conjoined twins throughout your life, some with only legs to move everybody, others with only mouths to ingest food, others with enlarged hearts to circulate the shared blood, and others fully dedicated to the sexual production of new offspring colonies."

Clearly each part of the colony does something the others can't, which is also true of bee colonies in which queens perform different functions from workers. But the bees don't have identical DNA, nor are they joined. The siphonophore is an intriguing evolutionary phenomenon. Scientists hypothesize that specialization occurred slowly over time, gastrozooids for instance gradually losing their venom while dactylozooids forgot how to digest food.

But do different parts of *Physalia physalis* communicate with one another? Can certain parts go rogue since there's no single brain in command? What would happen if the so-called goals of the four somehow differed, if conflicts arose, if one discrete part developed cancer? Would mutations in one zooid affect the others? And what if the hereditary information shifted in a group of cells—something called germ line segregation—would offspring not be wired to stick with the group? The Portuguese man-of-war, however fierce, is fragile. Nets disintegrate its tentacles; safe capture is problematic; the biological model is foreign to us and difficult to study.

——

An analogous corporate structure, also largely unseen by most of us, affects us vitally and essentially in every waking moment. Microchips have evolved into the very heartbeat of the globe, and we have allotted them enormous power. As with the survival of *Physalia physalis*, the ongoing production of this vital component of phones, cars, software, and data processing is highly specialized. Apple and Microsoft design the chips in the U.S.; a company in the Netherlands makes the machines that make the chips; and Taiwan Semiconductor Manufacturing Company (TSMC) uses the Dutch machines, along with their own sophisticated technology, to produce the world's most advanced microchips. While the current setup runs efficiently, it's potentially disastrous. All three components are crucial in preventing political and economic chaos, and the loss of any one part would take years to replace.

Physalia physalis is perhaps more agile. If an octopus lops off a tentacle, the Portuguese man-of-war can regenerate it while others work overtime to compensate.

We can speculate that dactylozooids became more potent when they became the sole captors of prey, when the entire job (but only one job) rested on them, just as we could say the unparalleled precision and sophistication of TSMC is in part possible because it's not distracted by designing the product.

What sort of evolution occurs in a human union, a marriage, a mini colony? Does each member become a greater or lesser being by virtue of symbiosis? As Virginia Woolf's Mrs. Dalloway wanders a London street toward the flower shop where she will buy delphiniums and roses and lilacs for her party, she muses, pictures wistfully a person she might have been ("Oh if she could have had her life over again!") before marriage, before children. She might have been like Lady Bexborough, "rather large; interested in politics like a man"; instead, she has a sense of being "invisible, unseen; unknown," as if her life had subsumed (like those smothering gastrozooids) some unquantifiable portion of the interests, aims, and dreams that characterized Clarissa as a child and adolescent, pre-marriage to a man of business. Now she progresses solemnly "with the rest of them, up Bond Street, this being Mrs. Dalloway; not even Clarissa any more; this being Mrs. Richard Dalloway." But she will have her party.

———

Specialization in a marriage, if ontologically suspect, has practical benefits. Clear divisions of labor can lead to smooth sailing, or the appearance
of it, like that azure bubble drifting along day to day in relative calm. In
my own outdated model, husband went to the office, wife raised the kids.
This allowed him a caveat for tough questions: "Don't ask me, ask your
mother, she knows the rules." It allowed me the autonomy of decision
making on such essentials as food gathering, breast feeding, sleep rituals,
etc. without agonizing discussions with him while in states of unprece-
dented fatigue. He did the taxes, I the laundry. He researched insurance
policies, I painted the house. For us such division saved time (tax audits,
spilled paint), skirted conflict between members of the unit, and possibly enhanced our awareness of the difficulty of functioning without the
other. I could not begin to reboot the internet or find the button on the
remote to switch HDMI channels when the TV goes rogue at 11 p.m. in
the middle of *White Lotus*. Yes, he could learn to transfer laundry from
washing machine to dryer before things start to mold, but it would be
inconvenient.

I grew up with the feminist movement. As a teenager I would never
have thought (if I thought about marriage at all) that mine would result
in such '50s-style roles. Even though motherhood is more glorified now
than it was when I quit my job to stay home, today's millennials struggle (at times with the futility of *Physalia physalis* in a headwind) to stay
impeccably on course with both motherhood and career. They're desper-
ately trying, and I applaud them, just as I do young fathers who are more
involved, who go to OBGYN appointments with their pregnant wives,
get up at night and give infants a bottle. But any young woman in this

position I talk to just sighs and shrugs. Who orders diapers and bottles in bed at midnight, remembers Teacher Appreciation Day, remembers to buy eggs, sets up pediatrician appointments, checks the supply of wipes, milk, packs the lunch, and makes to-do lists of stuff integral to effectual symbiosis? Fatigue and distress and outright hostility might come not from the difficulty of any of these tasks but from finding oneself in a murky no-man's land where one does them by default when someone else could but didn't realize that it fell in their domain or that it needed doing at all.

It's telling that only 10% of Fortune 500 CEOs are female. Does the specialization to get there demand too great a sacrifice? Clearly many women think so. Or is the problem the environment in which they strive? The lack of infrastructure to support perpetuating the species in viable ways? Or management distrust of anyone with small kids who might contract a raging virus, ear infection, or worse? My daughter, who has two girls under the age of four, heads development for a nonprofit that launches new musical talent. Her boss (female) recently remarked that she'd be hiring a candidate (male) to take on a capital campaign and praised his qualifications as: ambitious, hungry, without children.

＝

Although the Portuguese man-of-war is a highly functioning individual with efficient division of labor among its parts, it too is subject to conditions beyond its control: storms, ever heavier winds over warming oceans, wind-driven currents and coastal swells. Greater mass die-offs than what I witnessed on a Florida beach would occur without a simple evolutionary modification. The baubles of air can be either left- or right-handed depending on the attachment point of the tentacles. Each type sails at a 45-degree angle to the wind, but a right-handed man-of-war sails to the left of the wind, while a left-handed one sails to the right. When a storm comes up and winds drive a cluster toward the shore, about half will land, stranded, while the other half sails elsewhere, dactylozooids awake and armed.

A black-and-white fish, typically about a foot long, often swims among these tentacles, nibbling at them or at food bits the gastrozooids drop. While a lot more resistant to toxins than most fish, this one, called a man-of-war fish (*Nomeus gronovii*), still prefers not to be stung. With incredible agility, possibly due to forty-one vertebrae in its back, it dodges the dactylozooids' myriad filaments, or tries to. The Portuguese man-of-war tolerates the intruder because it attracts other fish attempting to prey on it, which the siphonophore stuns and eats. It's hard to imagine a human conceiving of this sort of symbiosis — maybe we're averse to risk or lack imagination. Populations of men-of-war are thriving and spreading north, capitalizing on climate change and even on the garbage we dump their way. Loggerhead sea turtles eat men-of-war because they have evolved to tolerate the venom. But they have not evolved sufficiently to differentiate between a pneumatophore and a lethal plastic bag. Fortuitous for *Physalia physalis*.

I didn't swim in the turquoise water off the coast of Florida. It didn't seem worth it, but for a week I walked the beach and dodged the Portuguese men-of-war, wondering if each was one or four, and how I could tell where one part ended and another began, and what parts were invisible or had been washed away. And I wondered if my husband and I were one or two, different as we are, or a single unit of mingled energy which, when one is lost, will leave the other feeling like less than one.

Beating it to the Sea

O N MY DEAD END street a neighbor posted three green signs depicting the profile of a generic turtle beneath the words "Caution Turtle Crossing." Despite that warning and "Slow Down to Protect Wildlife" and "No Outlet," motorcycles accelerate and a neighborhood teenager revs his Range Rover. I've rescued box turtles from the middle of the road and come too late for a few babies, just three or four inches in diameter, flattened nearly beyond recognition.

On the beach at the end of the road, tucked under grass where the woods meet the sand, Ella and I found the largest turtle I'd seen in the area. We were stunned. How lucky we were. Its shell was about a foot long, reddish brown, its feet and head mottled like an irregular pattern of mosaics sealed by grout. Although we moved gingerly, I fully expected it to shrink into its shell as any box turtle would. But it stared at us impassively and began a deliberate shuffle down the sand. "It's a sea turtle!" I gasped. Ella and I followed behind and watched its feet meet the bay water, its head duck out of sight, and body alight as if lifted by wings.

When Ella told the story later to the rest of the family, she said she had taken a piece of plastic off its body. I don't doubt her veracity, but I don't recall it happening. If it didn't, the detail was apt and prescient given the fate of sea turtles navigating our oceans. She felt involved in its survival.

The most common sea turtles off the shores of Long Island are loggerheads, which until that day I thought lived only in very warm water like the Caribbean. While snorkeling there some years prior, I saw adult loggerheads and swam quite close, near enough to see neck wrinkles and flipper nails, a yellow head and plated back where burgundy splashes made sunsets on each row of scutes. The turtle sailed downward, its crescent front flippers gently waving while the back remained still, a huge ancient creature unhurried and graceful in a milieu of flashing barracuda and skittish squid. Sea turtles have not changed much in the past one hundred million years of evolution.

A female loggerhead mates at sea at about the age of thirty. Then she trudges up a beach, digs a hole with her back flippers and deposits at least a hundred eggs like a bucket of golf balls, which she carefully conceals. Early one morning on a Caribbean beach, I watched a loggerhead get up from her crater and lumber toward the sea, her flippers pushing and heaving, leaving regular ridges in the sand like tire tracks of an SUV. She didn't look back. Her eggs might well be devoured by a raccoon, skunk, or fox.

Two months later under the cover of night, any surviving hatchlings would appear and aim for a glint of moonlight or starlight on the water. They would clamber frantically, sometimes bumping headlong into broken bottles and plastic or veering off track, distracted by lights onshore, as they run from frigates and gulls. Ahead lies a mirage of safety where a barracuda might snatch it after all, or a fishing net. As I picture the hazardous start of a turtle's life, I recall rocking my infant children and feeling their fingers around one of mine. I can't help but sentimentalize the turtle's plight—and question if that is so wrong. Would I do so if human interference had not amped up the appalling mortality rate of hatchlings?

If and when baby turtles enter the shallows, they remain prime targets, unable to dive, unable to take on a torpedoing frigate, and—with top swimming speed a half mile per hour—unable to outswim a grouper. What they have going for them is an innate knowledge of shifting magnetic fields across the globe, coupled with an imprint, perhaps a distinctive smell, from the natal site. This intuition, not a mother, acts as guide and the accuracy is astounding. Migration routes vary, but a hatchling, say from Florida, often starts in the Sargasso Sea, an area (between 20° to 35° N and 40° and 70° W) bounded by ocean currents where it feeds on crustaceans for about a year. Swimming with Gulf Stream current, it heads east and down the mid-Atlantic or paddles on to the coast of Portugal where the Gulf Stream divides, one part heading north to extremely cold, lethal water off Scandinavia, the other south to warm water along the western coast of Africa. At this juncture a particular magnetic field guides the turtle to turn south. A similar magnetic field simulated in a lab caused turtles who had never been anywhere near Portugal to make the same turns, suggesting that their ability to read magnetic clues is inherited and not learned. South of Cape Verde the currents shift westward and the migrating turtle heads back out into the Atlantic and finally back to its natal water precisely where it began. Loggerheads have also been tracked crossing the entire Pacific, swimming 6,000 miles from Baja to Japan, feeding, growing, then turning around and cruising a couple of miles an hour all the way back. Even after this lonely migration a turtle may not yet lay eggs, but complete another migration, spending twelve years before heading home and awaiting maturity there.

I think of the first twelve years of my children's lives, years of guiding, dispelling nightmares, bandaging cuts, cooking, hoping, and crying inwardly when they cried; the first school busses carrying them off and my trying to read their eyes when they came down the steps of the bus that carried them home. All those years the loggerhead swims and then, imagine—it arrives at its natal beach to find a seawall, a resort, its homeland vanished. Ancient Greeks believed that if a body went unburied, its spirit would wander the earth for eternity. This is the sea turtle's fate for without its Ithaca, it will keep roaming, searching, and never ever spawn.

A lovely bay with black sand beaches offered shelter to early Polynesian settlers of Hawaii. Far out in the bay they discovered a natural water spring and named the place Punalu'u, meaning diving spring. According to legend, divers with gourds collected fresh water from the spring until a time when they were stopped by violent storms. A beautiful green turtle named Honu-po'o-kea with a glowing white head crawled onto the sand one night and laid a sleek, dark egg. Her mate, Honu-'ea, with a handsome auburn shell, advanced onto the beach to help her dig a deep pool in the sand with a fresh water spring. Then they left. After some time the mother returned, and the baby hatched—she was the color of kauila wood, a wood so hard and dense that it sinks in water. Honu-po'o-kea remained while Kauila grew into a magical being who swam in the pool where the people now came for fresh water. When children ran about there and gathered shrimp, Kauila took human form and became a little girl who played with them and protected them from falling into the pool.

The legend departs from the natural life of sea turtles in telling ways. A male sea turtle never comes onto land as Honu'ea did to assist his mate. A female never returns to see her eggs hatch, nor to help a young one until it grows up. Sea turtles, with their tough necks and implacable eyes, protruding jaws and armored, barnacled backs, paradoxically evoke benignity and wonder, which may have inspired this legend and many others that depict them as strong and helpful to humans. And perhaps the allure goes further. One might expect to see these solid, ancient creatures on the ocean floor, not gliding on angel wings through beams of light, transcending their weight like birds, as humans have longed to do.

Such was the wonder that Ella and I felt that day when we discovered a small sea turtle on our beach. With no sense of irony or urgency, she (then four) shared a story that reversed the roles of the protector and the vulnerable as if it were the most natural thing in the world.

The Nimble Cuttlefish

The sound of the trumpets died away and Orlando stood stark naked. No human being, since the world began, has ever looked more ravishing. His form combined in one the strength of a man and a woman's grace.

ORLANDO, VIRGINIA WOOLF

AN OLD FRIEND AND I were walking around my neighborhood the other day, catching up on the lives of our kids and feasible shortcuts in dinner prep for husbands. "Remember when Oliver came to your house for dinner in pink high heels," she laughed. "I think he was four." Her son was now nearing twenty-five.

To be honest, I didn't remember. "Hmm, vaguely, but I know he always used to play dolls with Sophie and Annie."

It had been a good ten years since we'd broached the subject of Oliver's sexuality. When he was in high school, my friend said, "Yes, I mean, I think he's gay. It's looking that way."

By now I thought she'd know for sure, so I simply asked. She laughed her warm infectious laugh, possibly embarrassed, "I really don't know." She looked down the street toward the water, dappled gray between oak trunks and heavy foliage.

I thought for a moment about her uncertainty, remembering another friend who told me that sometimes she felt like a guy, sometimes a

57

girl—that everyone has some of each. More than a century ago, Virginia Woolf was bold in her expression of such an idea. Prescient and playful about sexual identity in *Orlando,* she wrote, "Different though the sexes are, they intermix. In every human being a vacillation from one sex to the other takes place, and often it is only the clothes that keep the male or female likeness, while underneath the sex is the very opposite of what it is above."

There's a little red fish that lives in the Caribbean. It's about three inches long and feeds along coral reefs like many colorful fish, but this one, the chalk bass (*Serranus tortugarum*) changes sex up to twenty times a day. It has both male and female gametes and optimizes its chances of passing on genes by being able both to lay eggs and fertilize them, though not its own. Typically, a chalk bass spawns two parcels of eggs and then switches to male so the burden of the next egg spawning that day will fall to its mate, who has by then fertilized the eggs. This system of "egg parceling" doesn't necessarily preclude the idea of some other male getting into the act, especially since the chalk bass live in large social groups, but the fact is very few do. To them it's obvious: division of labor and reciprocity inhibit cheating.

In the depths of earth's oceans lives an eel-like fish with tough scales and scores of sharp teeth on its protruding lower jaw, upper jaw, and tongue. This deepsea lizardfish (*Bathysaur ferox)* waits in ambush on the sea floor and lurches at passing crustaceans and squid, snatching them with teeth hinged backward to prevent any glimmer of escape. Scientists have found them in the abyssal waters off eastern Australia, the deepest sea region in the world, as well as off the coast of South Africa, New Zealand, South America, and across the Atlantic. It is perpetually lightless at these depths, temperatures stay in the 30s, food and mates are scarce. When a lizardfish meets one of its own, it has to be mating material—there's no going on OkCupid or Plenty of Fish—and nature has obliged. Each fish is equipped to be male or female at any given moment, but hideous as they look to us, they probably don't self-fertilize. Simultaneous hermaphrodite meets simultaneous hermaphrodite, and life continues, each taking one role or another in the quiet dark.

About midway through Woolf's novel, indolent, passionate, saucy Lord Orlando of Elizabethan England switches gender like a sequentially hermaphroditic fish. "Orlando had become a woman—there is no denying it. But in every other respect, Orlando remained precisely as he had been." With fish, the shift of gender can happen in either direction. About two percent of fish experience both sides in the course of their lives, these species spanning nine orders and twenty families, which means hermaphroditism developed independently in a variety of circumstances. I was snorkeling in the Caribbean when a guide casually told me that parrotfish, which I'd often seen poking among reefs, looking bloatedly awkward and oddly disparate in color—some flecked brown and white, others sporting turquoise and yellow—begin their lives as females (yes, the dull ones) and may turn into males (the brilliantly colored ones). Whether or not they feel a change in identity is hard to say, but males and females alike scrape algae off coral reefs and shroud themselves in a mucous cocoon at night. Then why the change? Larger males have a better chance at parenthood than do small ones. So when a parrotfish is small, it makes sense to be female, and some keep their gender, especially if they're sexually active. But if there's a shortage of males, females may make the change. Some species live in harems of forty females overseen by a supermale. If he dies, a large female will don the turquoise splendor, perform a coup d'etat, and assume his role, with no questions asked about origins or glass ceilings to break.

While Orlando herself shows no surprise at her change, society condemns it as against nature and makes a concerted effort to prove either that she is a woman, having been born that way, or that she is definitively a man. Woolf dismisses such pedantry with a wave of her pen, remarking that scientists can quibble over the distinction but the fact is Orlando is a man till the age of thirty and thereafter, a woman. It is Orlando who makes the choice. It is Orlando who expresses "rather more openly than usual—openness indeed was the soul of her nature—something that happens to most people without being thus plainly expressed."

The ocean conceals from most of us just how commonplace, vital, and natural gender flexibility is for thousands of species. In a reversal of the parrotfish, clownfish may start as male and wind up female. Bright orange with broad white stripes, they live in the Pacific and Indian Oceans in hierarchical groups in which only the two largest, most dominant fish mate. The male chases after the female, clawing, biting, and herding her toward his nest on a rock, perhaps near a sea anemone, which is deadly to most other species and thus a safe haven for the female's eggs. The male, mollified it seems, now protects them as do diligent male penguins who sit on the precious egg. But it's a short stint for the clownfish, a mere five days till the little ones hatch. If the female dies or disappears, this dominant male will become female and some other large male on the totem pole will rise up to take his place. This fluidity makes sense, though size for guys still matters.

Cuttlefish, which are not fish but very smart mollusks, use a sophisticated illusion of gender identity to ensure perpetuation of the paternal line. They are fish-like with small fins, but octopus-like with eight suckered arms reaching straight from the mouth and two tiny tentacles tucked underneath, which dart out to grab prey. With pulsating, rippling stripes along his sides, a male approaches a female with one thing on his mind. As in the human arena, one male's interest may spark that of another and enhance the value of the female. When a second male looms up, the skin cells on the side of the original male facing the competition quickly shift and squeeze to change color and pattern, creating a mottled look typical of females. The newcomer lingers—perplexed, deceived, totally excited about his courtship of *two* females—while the first male whose other side remains as it was—burning orange, striped, and suave—avoids a fight and fulfills his task. The second male slinks away, out of luck. The first cuttlefish is clearly aware of social context and savvy about his appearance and the reactions he can stir up. The female might have cooled if she'd seen his female side, but for better or worse, he keeps it hidden. Or maybe she wouldn't care, so tender is his lovemaking. Facing each other, head-to-head, the male embraces the female's head with his arms and remains still for a few minutes; then with his fourth arm he inserts a packet of sperm in a pocket beneath her beak, which he opens with a few movements, and departs.

The light show flanks of the cuttlefish quickly dim. All that intelligence, all those nimble skin cells, all those blazing reds and golds and browns lose luminescence after a single year. Life spans are often determined by external risks, and for cephalopods, there are many. Before dinosaurs strode the earth, many cephalopods slowly lost their protective shells. While cuttlefish retained a portion, the armor became an internal gas-filled chamber shaped like a surfboard, providing not defense but a degree of buoyancy. The body of a cuttlefish is squishy, its skin soft and vulnerable. Mating can't wait until a cuttlefish grows up. It has to happen in a flash, multiple times, in a single breeding season. At one year old, when human babies are clambering to their feet and nowhere near probing their gender identity or questioning the legitimacy of others', the cuttlefish is aging, its skin turning white in unsightly blotches, and with dying color come futile attempts to mate, to create camouflage, to wear bright things.

In the dark of the ocean cephalopods and fish go about practical, beautiful, often endearing ways of propagating the species. A mottled parrotfish sheds brown and dresses in sunshine, a little red fish lays eggs and fertilizes others, a clownfish opts to step down as king. And a cuttlefish flashes patterns of fiery red, olive, and maroon flecked with white signaling the electrochemical flurry of its inner life.

Gender fluidity happens gracefully and matter-of factly, with transformations as slippery as the intimations of otherness we sometimes feel and let go of. When my friend and I stood on the street that day, remembering little Oliver, wondering about Oliver as a young man, his mother admitting she really didn't know, I hesitated. A gust of wind bristled the leaves and a couple of deer bounded across the street as I held my breath.

"Other friends say I should just ask him," she said slowly.

"No! Why? Maybe he doesn't know what to say."

She pushed a strand of hair (that she let go gray a few years ago) out of her eyes, her dangly pearl earring looking suddenly elegant on the gray street under a cement sky. "Thank you," she said. "I agree."

Into That Good Night

AFTER A FEW HOURS of staring at hyperbolic adult smiles and toys dangled inches from her face, being burped and walked around the yard, three-month-old Ella cannot take life anymore. When her mono-syllables get tremulous and slight furrows thread her forehead and she can no longer look at me, I ease her into a BabyBjörn, make the straps snug, and head for the door. By now her wails are relentless. I tip her head slightly to put a pacifier between her lips, but she won't hold on, won't relent, though I hold it there while her screams grow thicker, angrier. Her eyes tighten, leaking tears, her face turns red as the ripe flesh of a plum. Her body is one flexed muscle as each exhale carries her frenzy up and down the street. I wiggle the pacifier against her tongue but it's all too petty for what is on her mind. She has to scream. I pick up my pace and she eases momentarily, taking in a shaky, staggered breath, letting go, and then another. She cries again, this one sadder, longer, this one from her throat, not her gut—she takes another stuttering breath, shudders, and sucks the pacifier as if it alone will sustain her, frantic little sucks at first, then the syncopation turns to steady eighth-note sighs and her eyes close. I keep walking, humming calypso tunes, listening to her even breaths, and hearing my own like a steady bass continuo beneath the padding of the Björn. One of her hands rests lightly on my arm, a crease at the wrist, the other scrunches under the sleeve of my T-shirt, her skin like a warm nectarine. Finally I sit down and lean back, her body rising and falling as I breathe. The neighbor starts up a leaf blower; I touch Ella's plump arm, look down at wisps of auburn hair. She doesn't stir, doesn't hear or feel. She needs to be without us.

But maybe she is not. In some form we enter her dreams, we along with trees, chairs, flowers, motorcycles, light, water, an open refrigerator with milk cartons and ketchup bottles, things that as yet have no name or function but hit her brain throughout the waking hours. Humans all need sleep to cope with the staggering maze of stimuli in the course of a day. Pelagic species, like bluefish and tuna, on the other hand, show no signs of resting because they swim in open water that is so monotonous they can process sensory input while awake. For us, the morass of known and unknown needs to slide from short- to long-term memory. The brain doesn't close shop during sleep but instead takes stock, reorganizes, stores, like arranging soup cans and boxes of cereal on a supermarket shelf, though our dreams are not so orderly in a way we might define order. We deceive ourselves in longing for sleep as a form of oblivion. It is survival, which Ella demands that I know.

———

While Ella's psychic and physical growth depends on rest, some species survive by staying awake. Giraffes sleep about four hours a day in quick ten-minute naps, usually standing up, or enter a half sleep with eyes slightly open, ears still twitching. In South Africa I watched a giraffe advance from a hilltop to the edge of a watering hole where he stopped and waited—waited as baboon families drank, followed by herds of zebras and kudu. He turned his head slowly, checking the horizon in all directions for excruciating hours before spreading his front legs wide and lowering his stately neck—risking attack by a lion, hyena, or leopard. Deer sleep only about three hours a day for the same reason, and ostriches sleep standing up, eyes wide open most of the time.

Dolphins have an enviable method of getting some rest while watching out. They slip into slow-wave sleep, or uni-hemispheric sleep, one side of the brain staying awake while the other half sleeps (like teenagers in an English class). The active portion keeps an eye out for predators and reminds the dolphin to breathe since they (and other marine mammals) must make a conscious decision when and where to surface. After the left hemisphere and right eye rest for about two hours, the dolphin switches sides. Female dolphins swim constantly for days after giving birth, slow-wave sleeping often, while the baby, who doesn't have enough body fat to be buoyant on its own, slipstreams behind and rises to breathe when she does. It will not sleep during the first month but shows no signs of stress, no rise in cortisol, none of Ella's fervid urgency for sleep and paradoxical resistance.

Despite air traffic, the sky also provides a low stimulus arena, allowing migratory birds to cover huge distances without sleep. Common swifts breed throughout Eurasia and migrate to Africa—usually *nonstop*. They spend ten months out of the year in flight, averaging twenty-five miles per hour and living till the age of about twenty, which means they fly the equivalent of seven round trips to the moon. The birds take wing in mid-July when the supply of insects in the north begins to thin and, rather than waste time on the ground, they live alongside their food source, aeroplankton, a mix of insects that blows in the air. Amassing a ball of it in the back of their throats, they either swallow or feed it to their young en route; they also copulate midair, like dragonflies. Because swifts rarely land, they scarcely have feet, their feet being too small for walking or perching to rest. Twice a day, they climb to altitudes of 10,000 feet and catch a half-hour nap gliding on thermals in their descent, trusting something other than an alarm to wake up.

Nearly three thousand years ago Homer wrote about the paradoxical relationship between sleep and survival. Odysseus is often compared to a lion, a species at the top of the food chain that has the luxury of sleeping a lot. And yet, when he dozes off tantalizingly close to Ithaca, his avaricious shipmates untie a bag of wind that blows him off course. He should have been a giraffe, watchful and untrusting. He then spends seven years beneath a shroud of near oblivion with Calypso (whose name means to veil or hide), ignoring his own identity and fate—*Odysseus* being linked to *odusomaii*, to feel pain and to inflict it. At last, on behalf of Zeus, Hermes arrives on Ogygia and tells Odysseus to wake up, get going, live out who you are.

Meanwhile, Penelope the loyal is a self-described insomniac: "When night falls and the world lies lost in sleep, I take to my bed, my heart throbbing, about to break, anxieties swarming, piercing—I may go mad with grief" (19. 582-4). In a cyclone of worry, words spinning round and round with nowhere to land, what did she murmur to the ghostly memory of her husband, what coy phrases to the marauding suitors? How bizarre that on the night when Odysseus thrashes around the main hall, clashing swords and slaughtering suitors, when men are groaning and dying, she claims never to have slept so well since the day he set out for Troy. This could not be coincidental. She is furious with the old nurse Eurycleia for waking her, for "interrupting my sleep, sweet sleep that held me, sealed my eyes just now" (23, 17-18). Nor is this Homer's attempt to heroize Odysseus and render Penelope useless; after all, he has made us admire her self-possession and fidelity for most of his tale.

Her deep sleep is sweet. It is not pure oblivion or ignorance but a reservoir where the being of one feels the being of the other like the movement of water after a body moves through, a slipstream where husband and wife sense the other and take comfort. Somewhere she knows he is doing what he needs to do because of who he is, as swifts trust the thermals on which they glide; in solitary wakefulness, he knows she is tranquil, safely away from the slaughter. In that nebulous medium the image of Odysseus interweaves wakefulness and rest, consciousness and unconsciousness as deftly as the threads on her loom create a shroud, for as Penelope remarks, "Odysseus. There was a man, or was he all a dream?" (19.363).

At this point she may be too world weary to trust her memory or her dreams, or too savvy to let him believe that she does. Restored by sleep, incisive, firm, and equipped to protect herself, she puts him to a test, a literal and physical one. She asks Odysseus to move the wedding bed that he built for her, knowing that only he would know it could not be moved without destroying the entire house. He is appalled by her lack of trust. His world and experiences have left him with more clean-cut though perhaps less accurate dichotomies. For her, the dream of a man woven through her days and nights might be the greater reality than the flesh and blood before her. Her ability to entertain both at once, a means of survival.

───

As Ella's wails soften, as she gives herself over to rest in the Björn, her arms go limp, her face presses against my chest—she who knows no concept of trust or fidelity. Does she hear the ragged bark of a dog down the road and let it melt into a mélange of images of this person or that, the rough cotton by her cheek, the ratchety sound of a rattle, the smell of my sweat, or does it remain outside and beyond the brief, illusory orb in which I try to protect her? Shhhhh, I whisper, shhhhhh. Sleep a little more.

Program Notes

O N A BEACH BORDERING Peconic Bay, a mid-size horseshoe
crab was nuzzled up behind a massive one, more than a foot across.
I found them on a hurried walk at 7 a.m. before my first Zoom meeting
that day. Every few minutes the smaller would push the larger, advancing
an inch or two before coming to a halt. Sand covered the front third of the
bigger one, concealing its eyes, while the smaller one occluded the back
quarter of the larger and its entire tail. Like an inverted bronze bowl, the
motionless one seemed a dead weight, nothing more. Advancing or not
didn't appear to matter. The smaller one is mourning the larger, returning
it to the bay, I thought, absurdly, and marveled at the effort and patience.
It can't leave the other stranded in powdery sand beyond the reach of the
tide, and so this labor. I assumed the large one was decades old to have
grown so weighty. I squatted next to the pair and waited. No one else on
the beach. The water was glassy, barely lipping the shoreline. Terns speared
here and there above the dune grass, and a few gulls stood watching the
shallows from boulders that emerged at low tide. Would the smaller one

make it to the water? How long would it take? Would the larger sink to the pebbly bottom only to be washed up and back when the wind picked up and tide shifted? The timing of the crabs' movement was erratic. I pushed back my meeting a half hour. When the pair moved, I couldn't see any legs in action, only an even inching down the sand, as if they were battery-operated, commanded by something elsewhere.

Keep going, I almost said aloud, as the smaller one stopped, leaving me as edgy as I get in a traffic jam when I don't know why or how long. Worse still, near the shoreline the traveling grew tougher as pebbles accumulated on the front of the big one and hillocks of rocks blocked the way. Five minutes till my meeting. They were inches from the waterline. Now, I thought, go! But the smaller one edged its tail side to side and turned 90 degrees, abruptly leaving the lifeless one alone, then, opening the hinge on its back, rummaged its way into the pebbles. It had seemed on the brink of closure after carrying the heft of the dead round and round. And I had felt *in* on something extraordinary while also *outside*, in limbo, unsure where the border fell between the two.

Left on the sand were figure eights stretching fifty feet across, patterns like Celtic symbols or the Nazca lines in Peru. The perimeters were clearly drawn, the interiors a basket weave of irregular loops, crisscrossing and overpasses, while the track to the water was nearly straight and uniform in width.

———

The Peruvian geoglyphs depict circles, lines running for miles, and about seventy animals and plants etched in the rocky, arid desert of the Rio Grande de Nazca river basin. To create them, the Nazca dug away about a foot of red oxide-coated pebbles to reveal the lighter sand below, at times outlining their figures this way, elsewhere clearing the interior of the design, in what must have been painstaking, throat-parching work.

Some archaeologists believe the figures were part of a ritual asking the gods for water. Others think the designs reflected the constellations or were some sort of calendar, like Stonehenge. Stretched across the desert, best seen from the sky though drawn thousands of years before the invention of the airplane, lie a whale, a long spider whose legs stretch straight in opposite directions, a mythical creature whose tongue hangs out, a monkey, and a humanoid. From the perspective of the ground, those who worked on the geoglyphs might have had no idea what figures they were creating, just as we with the technology to see the whole do not know why the lines exist.

The horseshoe crabs' rough serpentines were also inefficient, created it seemed without regard for utility or time. Circling and overlapping suggested ritual, a performance in nonhuman time, which I had missed, and whose vestigial notes I misread.

The larger crab I witnessed was a female, the instigator, the matriarch. The male had latched onto her back with small front claws and ridden up the beach to softer sand away from the water. She had nestled there and laid thousands of eggs, which he fertilized as she deposited them and moved on to possibly a second site or third to lay more. There had been a dance for new life, but I had concocted a story that inverted birth and death, influenced unconsciously by elements in my life, not the crabs'. Such imposing fictions, though not quite anthropomorphism, are probably common and warp our vision of wildlife we come across and know little about. The male had not abandoned his monolithic other to the currents and wind—as I had scattered my mother's ashes in the ocean— had not mourned the end of one older and more replete with genetic history and Paleozoic memory, but simply let go when his part was played. Why he held on till their return to the water, I don't know.

By evening that day high tide had covered their entry point to the bay, and tracks from a red pickup parked just a hundred yards down the beach had routed most of the design and ground into the sand where pockets of eggs must have lain buried. The horseshoe crab never knew, of course, how many eggs were crushed under the truck or dug up by a gull or a toddler with a pink pail, how many larvae hatched and made it to water, how many reached the first molt, shedding their papyrus shells in the shallows to be washed up, delicate and hollow. There was just a slow, eons-old act to be completed, and a persistent human need to concoct meaning from scratches in the sand under a narrow vault of the sky.

Two-year-old Ella takes a stick and draws lines and zigzags on the beach. She asks me to write her name. I say each letter aloud and she knows that these are the marks that make words that make the stories we read to her, which she inhabits and effortlessly commits to memory. I have shown her footprints, hers and mine and a gull's and a dog's. I wave at our shadows. She seems to accept evenly these parts of us in the sand, be it mimesis or the designs of a carefree hand or the record of a sprint to the water—the insubstantiality of shadow having no greater or lesser weight than letters written in bold caps. I think she feels satisfaction or affirmation or happy confusion when she glances at these marks in the sand or runs through them. We make castles and decorate them with shells. She eyes me with a sparkle and sticks her finger in the side, cracks it, closes her fist and crushes the walls. She relishes this power; it's easy, and condoned. More troubling are things that disappear and might or might not continue to exist without her, or she without them. She hides herself in full view as I pretend not to see her, casting my eyes about the sky and sea until they land on her and she beams. She has gone and returned. My

eyes have affirmed that she *is*. Now you hide, she says, keeping her eyes
open as I crouch behind a not too large rock.

Pebbles and jingle shells underfoot, stranded red seaweed, coarse sand. I
expect the grit under my heels, between my toes on any given day, knowing
that by the time Ella is my age this strip of beach where I firmly place
my feet, feeling permanence in the give of the sand and the daily shift in
configurations of tidepools and sand bars, might have vanished.

———

A few weeks later, Ella and I find a horseshoe crab at the edge of the
water, facing inland. That's a female, a girl, I tell her. Why is it a girl?
Because it's so big. What's it doing? I don't know. I don't know what she's
doing. Look, here are her eyes. Can I see it? Which means, Can I hold it?
She's shy, I say, but we can touch her back. You do it. Okay, but let's not
bother her. We wait. In a few minutes, the tail swishes ever so slightly.
Two chestnut pincers emerge about a half inch from the front of her shell.
She pushes back, picks up speed, opens and closes the hinge between her
prosoma and abdomen, propels herself out a few feet in the water then
turns to come back, rising as if she herself were a wave about to break on
the shore. She bucks up and down, suavely rotating to her back to reveal
all those legs, and flips back, seeming to meld with the slight movement
of the water. She rolls over again—they sometimes swim that way, they
have eyes underneath too, I say—but then she beaches herself flat on her
back and doesn't move.

What's she doing? Ella asks. I don't know. Resting maybe. Why? We
walk on to a stream and wade there, watching minnows and holding
fiddler crabs and avoiding enormous purple turds from two huge swans
that are often there, but not today. When we head home, I spot the
crab from fifty yards away, still in the same vulnerable position, flipped

over, crevices exposed, claws folded like the hands of a corpse arranged for a viewing.

I'm reminded of that morning when I first witnessed the serpentine tracks and presumed to read them. Now as I feel the slope of sand underfoot and wind around my arms, hear an insistent crow, see Ella stop to pick up a rock, the boundaries of that liminal space seem to vanish, and I recognize the obvious: the crabs and I had breathed the same air. A simple but critical fact. It took mystery, it took the extraordinary for me to understand that our sharing of space is not inconsistent with knowing that I walked on the sand for pleasure, they to mate, that I shielded my eyes from dawn streaking across the bay, while their ten eyes read ultraviolet and things invisible to me, that every element of this familiar place was perceived differently, and nothing weighed the difference.

The following day, Ella gamboled down the beach to the place where we'd touched the large horseshoe crab, but she was gone. Why? asked Ella. The tides. Why tides? Because the moon pulls the water up. She looked around the sky. But the moon isn't here. She was right, and I couldn't explain to her the disappearance of such things.

In 2018 a commercial truck driver ran through the Nazca geoglyphs, rutting the sand, as the red pickup had done, leaving long ravines and carving new insignia where others had lain for two thousand years. He had ignored warning signs and was arrested, but released since no one could prove ill intent.

EARTH

Your Ham is a Pig

WITH ELEVEN-MONTH-OLD ELLA IN a Björn, I walked to Peconic Bay along a road cutting through a forest of austere pitch pines, oaks, sassafras, hemlock, and wispy white pines. "Good morning, trees!" I chirped, and Ella, intent on picking up language ASAP, mimicked, "Morning, trees." "Morning, rocks." And so it went. Although the trees didn't return the greeting, they heard us and welcomed us—to her, their capacity for interchange went unquestioned. One day I walked into the kitchen where Ella, finger painting her high chair with yogurt, brightened and cried, "Tree!" I was touched by the conflation, attributing only positive qualities to trees. Her parents raised their eyebrows.

Just a year later we visited a local farm and fed a gentle old horse. Ella let him take an apple from her palm, feeling the velvet nose and whiskers, rubbing her hand on her shirt after, and marveling at the size of the bite—who eats what being of major interest to her. As we left, I called, "Goodbye, horses," and Ella, just two, piped, "He can't understand you."

Her logic startled me. And yet she tells elaborate stories, "Once upon a time, there was a little girl..." to her stuffed horse. So the divisions between humans and the animal and plant worlds are messy and ill proscribed for toddlers. Certainly, they are far more generous than we in allotting consciousness to things we readily dismiss. Ella finds two sticks or strips of bark or leaves and the party begins, one inviting the other over for cake and lemonade, or getting on an airplane for California. At that point, the tree has lost some of its tree-ness and become more essentially human in appetite, but her ease in handling, her immediate sharing as equals in conversations with a dead oak leaf intimates a kind of kinship and absence of hierarchy long lost to us. The sticks are not symbols; there is no such remove.

———

Ella bolts out the door and across the grass to check out my haphazard garden. I'm no expert on fertilizing or spraying for predators, but a pro at deadheading. I carry scissors. She wants that autonomy so I got her some blunt ones that fit her tiny hands. Once we found that my son's dog Lorenzo had bombarded a cluster of phlox, leaving a stalk nearly uprooted, the flower head toppled to the side, petals strewn like big wet snowflakes on the dirt. "That one looks sad," I muttered, giving it a trim. We continued on, snipping tiny daisies whose petals caved inward and salvia that had lost its vibrant purple blooms, leaving knobby stalks poking the air. Weeks later she ran out in the yard, eyed a snapdragon browning at the edges, affirmed it was sad, and dive-bombed with her scissors. I did a doubletake, wondering why I'd anthropomorphized the phlox, why I'd sentimentalized it. I had not told her it was dead, partly because it wasn't, but more fundamentally because I wanted to skirt the death conversation. Now I had conflated death and sorrow rather than just telling her this is the end—this sunflower, this columbine is dead— and it's how it should be, or how it just is. When my son was two, my

Siberian husky died, a sleek silver dog I picked up in a trailer park in
Alaska when I was eighteen. She was lean, dark-eyed, and elegant, part
wolf I'm convinced, and Paul loved her. "Where's Shantih? he asked.
"Well," I said, scrambling, "You know how the leaves fall off the trees in
winter, the flowers wilt"— this was lame—"and die?" Without missing a
beat, he looked up and asked, "And you?"

I lied, of course.

More often the dichotomies that surface for Ella in reading about or
being in the natural world are less abstract, less profound than life and
death—safe versus not, for instance, pleasant to have around versus not.
She pokes her fingers down the tiny holes of fiddler crabs and I hold my
breath. She finds a marbled salamander (which I've never seen before)
and I tell her instinctually not to touch it, fearing something, I'm not
sure, that it will bite her, that its skin will cause a rash? (Both absurd.)
Too late I know I should've said, the salamander is shy and doesn't want
to be touched.

A fly pirouettes around a bowl of peaches, alights on Ella's cheddar
cheese. I gently shoosh it away with a sweep of my arm and wait for her
to run into another room before grabbing the fly swatter and smashing
it. If a moth is huffing at the screen to get out or a June bug climbing the
stairs, I find a tissue and usher it out. But this summer Ella found me on a
chair outside my bedroom with baking soda plasters on both hands and all
down one leg. I had compulsively grabbed some crabgrass growing in the
garden and had struck a yellowjacket nest. Ping ping, one after another, the
stings hit my wrists and the back of my hands. I raced inside and pushed
up my sleeves, alarming a wasp trapped inside that stung my forearm while
another found my calf as I pulled up a pant leg. Ella didn't witness the
drama, but she knew I'd been stung. I told her I had interfered with the

wasps' home. Wouldn't she want to protect her house if a giant scooped it up? I don't know if she bought it as she stood there staring stony faced at the messy white goo on my arm and bags of ice drip dripping beside me. She now knows that bees and wasps can sting; I don't know whether she assigns mal-intent.

But she's fascinated by the possibility. In one of the *Babar* books, little Alexander falls into a pond, and we get an underwater view of a crocodile coming his way. Frantic, Babar launches a boat to save his son and hurls an anchor at the crocodile, which lodges in its mouth like a dentist's gag, keeping its jaws wide and useless. The crocodile thrashes about violently. I skip a few lines to downplay its fury. "Does he want to grab him?" asks Ella. "Why? What is Babar doing?" These are tough questions. "Babar wants to keep Alexander safe. He can't swim." This won't do—she knows there's more to it. As we finish the book, Ella demands, "Again." But she flips fast through the first third of the book and finds the crocodile page and just stares. Is it time to say, yes, some animals will eat you? You eat animals? Your ham is a pig? I can't do this yet.

How long can I sustain the ideas that trees have language, that animals have territory with legitimate boundaries, that we haven't the right to cuddle or malign any creature on sight because we are not it, that rose bushes and cacti, bees and leopards can hurt you without being inherently evil, that certain deaths are more finite than others? Or is the question, how long will such ideas interest her? Will they nest somewhere in her consciousness so that one day when she learns about the intricate network of fungi by which trees of different species do in fact communicate, she will listen without a skeptical shrug, will go a step further and learn how such messaging enables the trees to survive? That is my hope. Today when I tell her to close her eyes and listen to the wind exhaling and inhaling or a flurry of chickadees just above or the lap and furl of waves, she's into it. Today when I hear a white-throated sparrow chiming somewhere, she tries to locate it among the heavy oak limbs. When I push her on a swing, she says, "The wind feels my face." And once, when we clipped off a bunch of dead lychnis, clematis, and salvia, she gathered them into a bouquet and quietly planted them in the garden, patting the dirt around the stems and saying, "It will grow."

Ella's attraction to anything in the natural world isn't governed by functionality or a cultivated aesthetic, not yet. On the beach lie piles of Long Island grit—yellow and white quartz ground smooth and round, light oranges and pearly creams. Underwater, brightened by reflection, they shine. Pale pink eggs, little full moons with shadows, stones you could roll around in your pocket or put on your mantlepiece. "Here's a good one," I coo, offering Ella an ochre globe, subdued and natural and classy as the walls of a Venetian villa. But she turns away indifferently, reaches into the water, and pulls out an ungainly gray rock, heavy, jagged, and coated in slippery algae. "A big one," she remarks, plopping it back with no interest in keeping or collecting. She pries at a bit of brick buried in the sand and another hefty piece of conglomerate. She grapples it with two hands and offers it to me. "Here, you can have it!" "Thank you!" I exclaim, taking hold, shuddering at my banal taste in prettiness and its implications, the favoring of white and pink and yellow over darker shades, the gut preference for what appears polished and uniform, the arbitrariness of this implicit value system. What the hell am I teaching her?

≡

Cardboard books for babies and word books for toddlers show jungle animals and farm animals, all of similar size and often alone on a page. Kids learn cow, pig, duck, zebra, and so on. Often the animals frolic, tire, and fall asleep, as every mom wants her kid to do. But two issues occur to me: a rhino appears the same size as a mouse; a rhino might well live in the same place as a cow. We don't learn scale or context. Relative size is tricky, as I've learned when Ella wants me to climb into a niche in a tree with her, or try on one of her dresses. But where something lives and what lives with it is crucial for all of us. Yes, boy finds puppy or persuades parents to buy puppy and all live happily ever after. Then there is the plover that picks meat from between the teeth of a crocodile. Why, asks Ella, why? Because it finds food there. Why? Because the crocodile has been eating and the plover needs to eat too. Why are there jaws?

This scene appears in one of the most unusual books I've come across for kids, *Little Elephant's Walk*. Author Adrienne Kennaway spent most of her life in Kenya where East African wildlife impacted her deeply. Her picture book offers vibrant paintings of agama lizards, hyraxes, a genet, potto, aardwolf—animals I don't know, Ella doesn't know, and neither of us will remember, but the reasons for connectedness and modes of connectedness might create a sort of patterning on which other relations among plants and animals rely. Why are the impala running from the lion, why is an egret on the back of a rhino, why does a bird called a honey guide lead a ratel to a beehive in a tree stump, why is the crowned crane dancing? A leopard lies in a cool breeze high in a tree while a rattlesnake coils around rocks in a hot sun. These animals want something, need something; Ella wants to eat my sandwich, stampedes through the living room, needs her mom. We see a baboon with a baby on its back, a fruit bat enfolding its baby in its wings.

The growth of Ella's awareness of what these animals are up to depends somewhat on timing and necessity. Right now, no leopard lopes through my yard so I don't need to tell her they're dangerous. A snake might, but most are harmless, though not all, so the characterization becomes more complex, just as it already has with fleshy mushrooms in the shade, which she fingers and mauls while I inform her that you can eat some mushrooms but not others. Why, why—the persistence with which she asks leaves me at times facing my own ignorance, at times abruptly awake to the choppiness of my answers. Ella has probably seen enough pictures of porcupines to know not to touch one, but I'm uncertain how well a book illustration transfers to actual encounter. (She never conceived the size of a horse before meeting one.) Facts to be doled out about the natural world feel like ingredients in a sauce recipe that will result, maybe, in a mélange of fear and respect and affection, optimally at different times and in ways not mutually exclusive.

More critical than facts is how to prolong the idea of agency, how to reconfigure an entrenched hierarchy that drives a human perspective to override that of anything else. This is not mawkish empathy but necessity, a recognition of essential symbiosis, a plover in the jaws of a crocodile.

And someday, however painful, we will need to remind her of the salamander—people may touch you in ways you don't want—or what I should have said about the salamander.

Down a Dark Hole

WHEN I LEARNED MY husband was battling voles, I asked, "What's that?"

"Google it," was the reply, his go-to for anything from fluke recipes to methods of tearing up planks on a deck. Rather than learning about the creature's lifestyle, I found sites promoting its destruction, a vole holocaust, companies waxing lyrical about a looming apocalypse for voles and stoking proprietary fervor to save one's land. I scrolled through "How to Exterminate Voles," "Guaranteed Vole Control," "Moles and Voles—Exodus Exterminating," "Say Goodbye to Moles and Voles Forever!" And then I found pictures.

The miscreant is not quite a mole, whose eyes and ears are hidden to keep out dirt as it tunnels underground, and not quite a field mouse, whose body is slimmer, tail a bit longer. Still, voles look like Beatrix Potter's Mrs. Tittlemouse or Hunca Munca, gentle inhabitants of downy British meadows that occupied my imagination as a child. Rather than invading and destroying, Mrs. Tittlemouse guards against intrusive spiders and beetles and struggles to tidy her home. I can still picture Hunca Munca, modeled on a mouse Potter rescued from her cousin's cage trap in Gloucestershire, holding her adorable baby beside a purloined cradle. Despite some destructive behavior on the part of the protagonists, the tale is not an indictment of mice—or voles.

But now, like coyotes in suburban California, rats in Paris, deer ticks on Cape Cod, cockroaches in New York, voles have invaded and their social media profile has plummeted. They are the cause of two things that Americans abhor: trespassing and costly destruction of property. New rosebushes in my yard have turned skeletal, brown, and barbed, toppling at the touch of my hand. There are many children underground to feed. The meadow vole has a high-speed, high-voltage reproductive system. By a month old, a female vole has reached sexual maturity and might have ten litters of ten babies in a single year. If we dropped powder repellents down the vole holes that pock our yard, a few individuals might move, but hordes are ready to move in as vacancies occur. Any yard is a virtual farmers' market of grass and tree roots, bulbs and tubers. Two recent bumper crops of acorns here on Long Island have contributed to the voles' success.

As I walk around the yard, the ground gives way unexpectedly here and there, like a sigh, a sinking of spirits, a hint that the land (feeling more fat than muscle) suddenly isn't what I expect, and then I picture the voles like little miners whose ceiling collapses causing avalanches and landslides. Seems both futile (on my part) and unfair (on theirs).

On the opposite coast in the redwood forests of Northern California and Oregon the white-footed vole is desirable, little known, elusive, and rare. Conservationists traipse through the woods with detection dogs hoping to find these little creatures since survival of the environment, they say, depends on knowledge of its inhabitants. Once the rescue dogs have sniffed out a vole residence, conservationists set a trap to capture the animal gently and safely, but a dog might spend weeks without a whiff. Apparently it's highly unusual for scientists not to know much about a mammal. In this case they know only that the vole eats alder, is endemic to the coastal coniferous forests of Northern California and Oregon, and may or may not be arboreal. Mystery and scarcity make this rodent everything its counterpart on the East Coast is not.

My husband Brian paces around the yard cursing the voles and blaming them for every brown patch of grass.

"Maybe it's because you fired the lawn guy?"

"Nope. It's voles."

"Maybe it's Lorenzo."

"Doubtful."

He is at war with this unseen enemy—whole colonies avidly nesting, procreating, nibbling leaf stems, and gnawing on tubers. Like us, they love potatoes. A vole will plant himself next to such a find and simply eat and eat.

One day I found Brian driving stakes into the ground at random spots across the yard. "Ultrasonic repellent," he said. "They send out high beeps that the voles don't like." That seemed humane. We waited the requisite four weeks for them to depart, but our voles enjoyed the chorus. More holes appeared alongside the hydrangea and beside the newly planted garden of phlox and marigolds and zinnias. Tunnels swelled and ran for yards at a time. "Okay, we'll stick a hose down the holes and flood them out," he vowed. We counted eighty-five holes and stopped counting. The whole yard would flood, he decided. Or turn to quicksand.

Horticulturalists suggest hanging screech owl nest boxes near your garden to save your crop, but my husband has not yet taken that approach.

Before I knew what a vole was, I spotted a mouse on the back of a lawn chair at about the same time that my son's dog Lorenzo did. There perched a pudgy ball with a short tail and ears like flower petals balancing on the top rung, and there stood Lorenzo, nose aquiver, eyes beaming, darting at the chair and positioning himself to do *something*—tease it, devour it, maul it, or toss it about like a play mouse—I didn't know and didn't want to witness. But I couldn't catch the dog, the mouse seemed paralyzed on its high-rise, and the dog is really fast, probably faster than a mouse, so what to do…I picked up the entire chair and flicked the mouse behind a pine tree thinking I might deter the dog for a split second, which would be long enough for the mouse to scurry and hide. Later, in a tumble from innocence due to my research online, I realized this cute brown and white brown-eyed thing I could cup in my palm was a vole, maybe the very vole that ate the rosebush.

"You *saved* the vole?"

"What would you have done?" I asked Brian as he turned his back.

"I'd see if it could've outwitted Lorenzo."

"Really?" I called. "A gladiator fight, great."

Prairie voles, unlike the promiscuous meadow ones, choose a mate for life and show powers of empathy usually attributed to great apes, crows, wolves, elephants, and some humans. At Emory University scientists separated pairs of prairie voles and subjected some of those they removed to loud noise and minor electric shock. When they were reunited, the voles spent a much longer time grooming the stressed voles than those that weren't stressed even though the group that stayed behind never witnessed the stress. A vole consoled only its mate and did so after the first test, so the response was not learned. Somehow the vole just *knew*.

Is it a leap to call prolonged licking empathetic? Is biological data more convincing than subjective inferences of emotions in a vole? Stressed and un-stressed voles had the same levels of stress hormone, meaning the one that remained free actually became anxious to the same degree as the one that had been caged and tortured. Biologically proven empathy. The observing vole even froze when it heard a tone associated with shock and groomed itself to the same degree it groomed its mate. Does this measured physiological response legitimize our calling it a feeling, a feeling of empathy more real, I think, than that of a human shaking his head for the person in a car wreck as he sits in traffic cursing the crash for making him late to lunch.

As reported in *The Atlantic*, Emory scientist Larry Young says there is "cognitive empathy," which leads one to imagine, what if that were me?—a projection of self that the person in the car is attempting. This is Atticus's lesson to Scout in *To Kill a Mockingbird*: true maturity occurs when you can put yourself in someone else's shoes. Clearly, the vole isn't reasoning and contemplating being some *other* vole. But then there is "emotional empathy," which according to Young is "more of a gut, instinctual feeling." That seems to underlie the vole's impulse to console, and I think it has a certain validity. When books on child rearing counseled me never to allow my baby in bed with me, I held out for months but finally succumbed when I felt (instinctively and beyond all reason) that my baby needed something. Whatever it was, she did, but not for long. The problem resolved.

On that afternoon with Lorenzo, my gut feeling was to save the mouse. Would I, had I known it was a vole that had cost me hundreds and eaten up my time? That very morning I had walked around with a bucket of topsoil filling holes, losing track after counting to eighty, and stomping down on the swollen soil. If the holes open up again, say the exterminators, you know they are still doorways to active residences, not shells that former tenements left vacant, and you know to pour poison down those holes. You can feed the voles chemicals that cause internal hemorrhaging. Or introduce snakes to your yard. Or do nothing.

As I picture voles and their arterial tunnels and graves, I imagine the
world turned inside out, the invisible made visible—worms, bats, ter-
mites, moles, burrowing toads, locust larvae, grubs, ants, bustling and
squirming and wriggling in the sunshine, the way a centipede does when
you overturn a rock, destroy its refuge, electrify it to take cover. I see a
sci-fi world of this earth we walk on, a world after Armageddon. All
that we call brown, unthinking brown, assumes nuanced shadows and
hues of walnut, amber, ecru, dun, hazel, chocolate, russet, umber, puce,
khaki, fawn and oak in milky morning light—or the glare of midday
or the streaks and pitches of sunset—just as all that scrambles through
our dreams or meanders through our semi-conscious thoughts at night
dresses differently by day.

Were the invisible made visible, the unseen seen, we would look for the
sacred in new places—a hefty taproot storing food and supporting a leafy
life above, the seed or the radicle or the searching absorbing networks of
fibrous systems strung like the lacework tendons of a suspension bridge or
the mirror image of tree branches and twigs that crisscross the sky. Maybe
the divine is not a worm itself but the worm's control of each segment of
muscle and setae as it elongates and shrinks, sets up anchor and knows
when to let go. There must be more permanence in a windless arena, more
certain survival than occurs in seasons where we witness visible birth and
death, unfurling buds and crinkly fingertips of ferns—mortality—which
is not in the usual definition of the divine.

Because of the voles and their destruction, I wonder what to kill and
what not to kill. When I've been away for a few weeks, I crank open a
window to find a mass of ants and an anthill on the ledge. The ants scurry
helter-skelter, I grab the Fantastik, let loose a cloudburst of chemicals, and

DOWN A DARK HOLE

whoosh, wipe away a homestead with a paper towel. Afternoon sunrays suddenly light a spider web strung between the lip of the window and a strand of ivy growing up the wall and a flowering hydrangea. It's twice the span of my hand, concentric, pleasingly symmetrical but not, and a small furled spider sits at the heart of it. The silk is pure protein; spiders need to eat part of their web to compensate for the loss of energy in making it; the silk is five times stronger than piano wire. Will such facts stay my hand, already raised to wipe it out? Or will the aesthetics, enhanced by the whimsical position of the sun that happened to light each deceptively ephemeral strand and its position in the entirety of the design? I'm dazzled by the web's size and complexity and the unknown quantity of time the spider was at work in my absence with every right to build here where I was not. The paper towel is dappled with dead ants. A mosquito quivers in the web, and I leave it there. I leave the spider its kingdom, knowing how shamefully arbitrary are my choices.

Whimsical. Capricious. Arbitrary. A lot of decisions feel that way these days. Do we put together a decapitation squad to off a contentious leader or wipe out a country just nanoseconds before being wiped out ourselves, maybe? I could read a rationale for either impulse. It has become easy to imagine a country in which all internet sites relating to an antagonist contain nothing but methods of annihilation. To kill or let live, don't overthink it. Feel free and safe not just this summer but for summers to come.

Recently my husband remarked, "I think the voles have moved. They're over at Larry's now." He says it's because he put nasty smelling powder in the holes by the roses, but I don't think so. There had been *hundreds* of holes, he didn't use *that* much powder. But I don't reveal my doubts about his conquering prowess. Instead, I picture new babies agog in a new land, little eyes opening to hearty hydrangea roots, and mothers growing plump on Larry's newly planted pines along his driveway and freshly installed andromeda populating all the earth not covered by lawn, and all those roots luxuriating in fresh earth turned over and loosened for easy vole access by a guy that Larry hired to beautify his property and fatten his profits. Old stands of oaks shield the vole holes from osprey that wing overhead, and Larry keeps his corgis on a leash for reasons that have nothing to do with burgeoning populations underground.

Chemical Warfare

S INCE THE SUN HAD made a decisive, long-awaited appearance, I should have been lying on a beach with a beer. Instead, I was sprawled on my gravel driveway, stones jabbing my left thigh, as I pulled up spidery weeds that had crept up through the stones and spread their tiny, leafed limbs here and there. Beneath the shade of pines, they were less conspicuous. Out in the open around the cars, they sprouted shoddily, as if one had failed to sweep a room. The name of this weed was unlovely and stubborn as well—prostrate knotweed—and the best solution for eliminating it was to subject it to vigorous perennial lawn species, but naturally I wasn't going to do that in a driveway, so I continued to pluck.

Scanning the property, I noted blotches of dirt, sprinkled with blue seeds, and newly sprouted lawn grass looking regal if somewhat fragile. Here I was tearing my cuticles pulling out green things while a mere four feet away we were striving against unknown odds to see green things appear. Brian had paid landscapers hundreds of dollars only to find broad swaths of crusty brown grass by mid-July; he had gone online and educated himself about which fertilizers kill moss, which don't kill your dog, optimal climate for seeding, unbeatable mix of peat moss and topsoil, best brands for arid soil, shade, full sun, variations as specialized as sneaker brands for walking, running, standing, tennis, kickboxing, basketball. I felt pretty absurd.

So what is a weed? By general consensus: "a plant in the wrong place." This anthropocentric, relative definition tells me exactly nothing about the 250,000 species of plants dubbed weeds. "Wrong"? By that standard should we say: a dog is an animal that might be in the right place; a rat is an animal that is in the wrong place just about any way you look at it unless you're a snake, owl, eagle, or weasel. If you happen to be a rhino, flamingo, wolf, polar bear, orangutan, kangaroo, or termite, then we humans are in the wrong place. Australia's Department of Environment and Energy defines a weed as "any plant that requires some form of action to reduce its effect on the economy, the environment, human health." I think we fit that one too.

Poking through the gravel, I also found "legitimate" grass sprouting here and there. I suppose some seeds had blown from the lawn. So those healthy tufts in the driveway were now out of place. Had grass become a weed? And what about the clover, also making a show? "There's clover and there's clover," said Brian sagely. One year masses of clover arose in the lawn, at first looking quite pretty and green, adding texture, blending nicely, till it kept growing, reaching heights of six inches, the leaves tilting arrogantly from long spiky stems held tenaciously in place by deep roots that required meticulous surgery with knives and skinny trowels to remove without totally scarring the surrounding lawn. That kind is on his shit list. This year he has given license to a smaller variety, Dutch white clover sprouting little globular flowers, which are fragrant and silky underfoot, appealing to bumblebees, and resemble a field of miniature daisies till the lawnmower decapitates them. Still, the clover soldiers on, bacteria in its root nodules producing nitrogen in the soil, which is a quintessential ingredient for any decent lawn. Other sources of nitrogen include manure, activated sewer sludge, compost teas, guano, ammonium nitrate, calcium nitrate, and synthetics like sulfur-coated urea, resin-coated urea, and isobutylidene diurea—which sound like the contents of an outhouse or an infectious disease, making clover look pretty good and effortless.

About a week later I faced a dandelion dilemma. In my ragged front yard, dandelions pop up regularly and profusely just as, in the aftermath of tornadoes and nor'easters, we're about to give up on the advent of spring. There they are, looking so yellow, so cheerful, so redolent of childhood when one made dandelion chains and winced at the sour taste of the stems. They distract the eye from crabgrass and goosegrass; they verge on the relief and sporadic joy of daffodil sightings on distant hills (though now those are planted in calculated clusters, neighbor competing with neighbor for density and numbers). Although I've never plucked dandelion leaves for dinner, I know they're rich in vitamins and antioxidants. After an inner smile at the return of the dandelions this year, I demurred, considering aesthetic and now social implications. Scores more appeared and I grew marginally embarrassed. I, who care less than would a chipmunk about the neighbors' opinion of my yard, shuddered slightly at the sight of my seemingly neglected lawn. I also rationalized that I should dig out the dandelions before they went to seed and lost their gold, turning to ghostly dust that would spread boundlessly when the next storm hit or when a child happened to blow the parachutes asunder. I proceeded to dig out a barrelful, leaving holes like so many vole holes scattered about the yard, only to learn later that dandelion roots can be three feet long and that any lingering fragment of a root has the power to regenerate, so the slaughter was in vain. Appearances are fleeting. But we Americans think short term.

As I stabbed at the dandelion roots, I fondly noted a buttercup nodding in the breeze. Delicate, shiny, tiny, turning the light of the sun to prophesy with its dim sheen under our chin. Growing there seemed as good a place as any, certainly not the *wrong* place. I don't have horses or sheep to whom the buttercup is toxic, though grazers usually have sense enough to stop eating it before it kills them. So even though the College of Agricultural Studies at Penn State says, "weeds are plants whose undesirable qualities outweigh their good points," in this case, *they* were wrong.

True to the American myth of infinite possibility—the sort of optimistic "you can be whatever you want to be" propaganda we feed our children—Emerson offered a kindly, near-mystical definition of weeds: "a plant whose virtues have yet to be discovered." The philosopher also wrote, "A portion of the truth, bright and sublime, lives in every moment in every mind" and "The highest revelation is that God is in every man." I'm glad he extended his optimism to weeds. It should give us pause. But these days the definition just seems cute and a little naïve, making me faintly nostalgic for a time that actually may have been no better than our own. I can't get past Emerson's refusal to admit our seemingly innate propensity for evil, and my husband can't stand by and wait for the virtues of crabgrass to appear to him in a transcendental vision. The other day he called me at work. "It's just... so sad," he sighed.

"What?!" I returned, fully alarmed. I had thought he was peacefully writing his books or replacing boards on the deck.

"The lawn."

"What about it?"

"All that work, all that seeding…."

"And?"

"The crabgrass is everywhere. And these other spiky weeds. Never saw them before. Everywhere." There followed a deep, fatalistic breath. "I don't even know what to try…I threw on Scott's fertilizer with pre-emergent killer, but there are three more steps, such a long process…." I waited for the prognosis. "It's… so, so …sad." Such despair from a man who sees the glass near full no matter what's in it, who calms our daughter when her boss chews her out, who believes "things will be okay."

I started laughing. "Brian, It's not chemical warfare in Syria."

"I know. But weeds. What the fuck is up with weeds?"

After we hung up, I thought, what the fuck is up with lawns? Pre-industrialization, Americans were content with dirt patches and cottage gardens outside their doors. And although Emerson advocated we look to America for its own culture rather than back to Europe, wealthy Americans had traveled to England in the 18th century, soaked up visions of expansive lawns on family estates, and wanted to import the phenomenon here, thus seeding the association of fine lawns with money. But temperatures are more extreme here than in England, and many attempts failed, allowing for an invasion of unruly species that would not have thrived in our virgin forests but got a foothold in semi-cultivated areas. In the opening pages of *The Scarlet Letter* Hawthorne facetiously mentions that as early as the 17th century the Puritan colony in Massachusetts had "a grass-plot, much overgrown with burdock, pig-weed, apple-pern, and such unsightly vegetation, which evidently found something congenial in the soil that had so early borne the black flower of civilized society, a prison." An inauspicious beginning for lawns and democracy. The common man in the next century didn't have the time or means to maintain a lawn; one needed a grounds keeper skillful enough to wield a scythe to trim the slender blades, which is hard even to picture. Presidents Washington and Jefferson used an alternate approach, inviting sheep to the White House to keep the lawns in check.

Post-Civil War many middle-class Americans left cities and launched the phenomenon of the suburb. Frederick Law Olmstead, perhaps best known as landscape architect of Manhattan's Central Park, designed a suburb outside of Chicago in which he prohibited walls and required each house to be set back thirty feet from the street. Lawns became the great equalizer, the connective tissue, the democratization of America. Now one could look down a street and enjoy the vista of not only one's own yard but also that of one's neighbors—a long, open, unifying carpet of green. Frank J. Scott, a contemporary of Olmstead, took it a step further in his book, *The Art of Beautifying Suburban Home Ground,* where he admonished that it was "selfish," even "unchristian," not to adhere to the suburban aesthetic, take collective responsibility, and maintain a plot of velvety verdure for the pleasure of all.

As might be expected, the U.S. Golf Association joined the grass bandwagon in the early 20th century, finding species that would flourish here, and the American Garden Club provided the official mandate and principles to which, now armed with hoses and hand mowers, we must aspire—the American yard should be "a plot with a single type of grass with no intruding weeds, kept mown at a height of an inch and a half, uniformly green, and neatly edged." Despite his barbaric yawp for democracy, Whitman might have staggered from his grave at such curbing of generative properties, trumpeting for grass that grows "in broad zones and narrow zones," grass that is "the flag of [his] disposition," which would be nothing if not unkempt, and grass arising from the "beautiful uncut hair of graves."

≡

I walk over lawns in East Hampton and find not a weed, nor a blade out of place. I see armies of men arrive in pick-up trucks each spring to make sure this is so. As of 2023 the U.S. supported a 153-billion-dollar lawn industry, but grass-defined democracy is on the decline. Population explosions of deer in communities along the eastern seaboard have led to massive fence building. Many middle- and upper-class Americans hide behind iron spikes, imprison themselves, and block out their neighbors to safeguard their peonies. Gone is the civic duty to mow your lawn. Now it's each to his own. Second, the .01% invites us to *imagine, not view* their English estate lawns adamantly concealed behind thirty-foot border walls of privet.

≡

Weeds can be invasive. Sometimes they come from other countries and threaten indigenous species, sparking legislation that targets some types over others. Weeds enjoy moving to new places to colonize.

Some weeds, including horseweed and pigweed, have garnered the distinction of being "superweeds," ably flexing their muscles and warding off herbicides. Scientists quibble with the term, though, since they say these species aren't inherently any stronger than others, they've just developed resistance to the popular glyphosate, patented as Roundup, and hundreds of weed species have figured out how to do this. One solution is to mix in other herbicides, which naturally is healthy for the ever-growing pesticide industry, or return to an older chemical called 2,4-D, which was used to strip forests in the Vietnam War and has been linked to cancer, kidney damage, and even the emergence of antibiotic-resistant bacteria. In either case, the idea is to fight resistance to herbicides by using *more* of them, sort of like an approach to the nuclear arms race.

To distract Brian from his weed misery, we went to a local nursery to buy flowers to put in pots for the deck and to scatter around the margins of the grass where things were looking especially scruffy. Why not add a daisy bush to a nest of violets (which happen to be weeds)—who's to know the difference? After some hours of deliberation and damage on the credit card, we came home with a station wagon full of flowers, ones I'd never planted before, like pink hibiscus from Florida, kudos coral, and dahlia. In a side garden where all that currently grows are tall grasses, I had the inspiration and, I thought, genius, to plant five yarrow plants whose sagebrush-gray leaves and stems blended perfectly, lending themselves to the semi-wild look, while the flat-top yellow blooms were eye-catching and cheerful, a fine distraction from the foxtail and moss now battling with the grass.

Later, leafing through my weed reference book, I happened upon the unexpected: yarrow. I'd dropped $50 on a weed. Should I tell Brian? My appreciation of the yarrow blooms laughing in the garden was not diminished one iota. Instead, I scanned the list of weeds for possible additions, relishing their names the way a puppy rolls in the grass after a bath: Bouncing Bet, Nimblewill, Curly Dock, Black Nightshade, Chicory, Cocklebur, Wild Madder, Moneywort, Fall Panic Grass, Shepherd's Purse, Poor Joe, Daisy Fleabane, Flower-of-an-Hour, Foxtail, Trumpet Creeper, Morning Glory, Vining Milkweed, and Nutsedge. They chime folk legend and oral histories. They sound as whimsical as a yo-yo, tangy as apple cider—fertile, affordable, and easy to grow. Velvety carpets of mowed grass, by contrast, harken to an illusory American aesthetic, which is too singular and too complacent. I reach for a beer, and consider.

Out the window Brian trudges by with his wheelbarrow loaded with fertilizer, step 2 in operation weed annihilation. He pauses, bends down to pluck at something, shakes his head, and scans the horizon for an augury as the tree tops undulate and crows startle, lurching suddenly into the sky.

White Ash

SINCE APRIL, I'VE BEEN watching a tree that won't die. Today, November gusts career along the north shoreline, shifting capriciously from the southwest to the north. One leaflet ripples at the top of the dying tree, flag-like, feather-like , telling me the tree is a white ash. Its compound leaves, silvery on the underside, are set in opposition, pinnately, a word whose root comes from the Latin *pinnatus* meaning feathered, and *pinna,* wing, which seems increasingly ironic as torn leaves and twigs litter the ground at my feet.

The ash lives on the margin of scrub oak woods that run up to Peconic Bay and halt at a stony strip of beach, which yawns at low tide, shrinks to thin-lipped impassivity when the moon pulls up the water twice a day. Even at high tide I don't find the tree immersed. But its roots belie that, blatantly. The tree has formed a ledge two feet above the stones, over a space devoid of soil and leaves, the trunk anchored only by roots behind it that reach back into the woods. To either side thin roots like cables stretch twenty feet or more, running alongside the bank in which they must have once lain concealed, messaging oak roots and sucking water that filtered through a clutter of fallen twigs and leaves and tumbled rocks above. A few brittle roots point over the low bank toward the water, short as parsnips, yellow and dry; another, longer, extends like a naked fibula with nubby distal end exposed. Frayed roots that dangle down snap off to the touch, but the bleached lines running side to side feel meaty and malleable like an old carrot.

The tree doesn't give me a hint if it will live for fifty days or fifty years. Only data on rising water levels documented by some scientist who has not seen this tree would know that. So I have an answer, maybe not *the* answer. The trunk is just a little tilted, its wily branches curl this way and that, smug in their bay-view location. Some of its branches must already be driftwood that bobbled across the bay to Connecticut or washed back up here. Maybe I have found a stick or two to throw for my dog.

How much exposure can the roots take? Each high tide nibbles at the eroding bank, and nor'easters, harsher and seasonally erratic as global conditions shift, tackle the shoreline. In the news I gaze at maps of whole continents rimmed in red where cities will no longer be in twenty years or fifty. The West Antarctic ice sheet is collapsing and now the East one, too. Glaciers crack and rumble into the Arctic Sea, and Hurricane Sandy pummeled water through midtown Manhattan not so long ago. I can't picture, identify, or fathom the number of species already extinct. I stop reading, knowing well the argument not to do that.

Adjacent to the tree is another whose fully torn-up roots splay like an asterisk, whose trunk rises a few feet, takes a ninety-degree turn inland, extends about ten feet horizontally, turns another ninety degrees toward the sky for sunlight, I guess, that it can't absorb. I've walked by it so many times and never looked at it until now, this gray-bearded man squatting on the sidewalk.

My spry mother used to say, *Don't let me wind up a zombie, slip me some arsenic.* But no. She with her white hair and weathered skin sat for years, not as free as driftwood, not as valuable as dead wood to a forest.

Along the pebbly coast hundreds of white shells crack easily underfoot as I pass the tree. The light sound, not the shape of the cupped little saucers, alerts me that I'm stomping on hundreds of baby crabs wiped clean by gulls, that the pathway of pale twigs resembling french fries is a mass grave of severed legs.

We clear dead trees, tuck old people in nursing homes. But dead trees can survive fires. Their hardened, charred limbs might endure for centuries. Living trees can rot and live; living trees can grow hollow and live to tell histories beyond themselves. The boundary is not as definitive as it is for me; I figure this land is it, this edge of the woods, this ledge.

Forest managers sometimes *cause* living trees to decay. Slice off the top. Girdle the tree with a chain saw. Inject it with a dowel infected with heart-rot fungi that decomposes the heartwood till it softens enough for a woodpecker to chip away a niche in which to nest. Or until the heartwood separates from the surrounding sapwood and sinks to the base of the trunk, leaving it hollow, open for rent to a raccoon or bear. Dead trees and downed wood provide food or shelter to a majority of wildlife species in the forest—spiders, ants, salamanders, rat snakes, tree frogs, flickers, owls, bats, bees, wasps, and more—at some point in their lives. And so I am wrong to find the process, this systematic execution, so vicious.

Here in the east, as well as in forests of western and northern U.S., little gray moths lay egg masses on the underside of the needles of hemlock, spruce, and Douglas fir. Each larva spins a cocoon and tucks itself in a cranny of the bark until spring when white-speckled caterpillars, budworm, begin to feed on old needles, pry into new growth, and make webbing on which to rappel down the tree in search of more. Zoom out, the infested forest looks burnt. Within, trunks soften enough to afford nesting sites for woodpeckers, which eat budworm. Eventually the tree dies and dead logs house ants, which also eat budworm, and woodpeckers eat ants.

Walt Whitman said to the grass: "It may be you transpire from the breasts of young men." His pre-scientific, optimistic cycle of life and death observation has some validity. Still, I'm tired of the adage—death is necessary for life—it has come to sound like a platitude to help one bear up. And to flip it, to say life is necessary for death seems obvious, that is, for death to occur. But for death to have *being?* Budworm within ant within woodpecker within tree makes death seem alive.

When my mother was cremated, I sat alone at home and pictured a metal gurney sliding her into a hole in a metal wall. Unlike a dead tree, her corpse would burn. But mothers live on in hollow spaces left behind—as if daughters were the vacant tree in which some turn of phrase, some admonition or exclamation took up residence. It's a strange reversal, mothers being the original carriers of life, the bearers of a full womb. A hollow space is not a vacuum, as a black hole is not empty, as dead wood is not without life. Seeds of birch and pine germinate quickly on dead limbs, which store carbon and provide nutrients for young saplings indiscriminately.

Stumps and fallen logs also hold down soil on which they grew and mitigate erosion, which humans with the means may go to lengths to prevent. Along the south shore where the Atlantic meets Long Island, trucks haul in massive boulders, and bulldozers stack them against impending surf and uncertainty, which threaten the mansions gazing out indignantly just above. Usually those waves look pretty, tossing their spindrift in the sun. Somewhere among them my mother's ashes swirl and dive, washing up on the tongue of a wave and drawing back.

Salt and wind assault the white ash today and will do so tomorrow and the next. Exposed strips of roots will shrivel and crack. Underground, the tree may already be passing on its stores of energy, willing away its goods. Underground, wispy filaments of fungi called ectomycorrhizae grow on roots and draw carbon and nutrients from trees; in return, they filter through the soil, expanding the range from which a tree can absorb minerals and water and enabling a system of forest-wide trade.

Researchers in British Columbia planted seedlings of ponderosa pine and Douglas fir and netted the roots so that they couldn't connect. They stripped the needles of the Douglas fir, sending it into distress. Mycorrhizae branching from the roots slipped through the netting, carried carbon, nitrogen, and water from the fir to the pine and an alarm that induced the pine to produce protective enzymes. Why was the fir so generous, so solicitous? Or was it the fungi? Or did stores of food simply move from high concentration to low? The fungi might have wanted to keep the pine healthy as a future source of survival. The fir might have passed on its stores of food figuring that offspring and relatives lived nearby, or that differences among species no longer mattered. If root or fungus intentionality seems unlikely—still, what happened, happened. The fir's problem was existential; its action, which happened to serve the interests of the forest ecosystem as a whole, was ontologically affirmative, nondiscriminatory, radical.

Humans in the U.S. have never been wired this way—from the social hierarchies John Winthrop brought to our shores, to laissez-faire economics rewarding killer instincts, to the individualism with which we arm (and often isolate) our children, to Emersonian self-reliance and the incessant focus on personal identity in our literature. While splintered into regional and political factions, we find that beneath our feet massive networks of mycorrhizae sustain not only trees but also chaparral, prairies, Arctic tundra, and grass (Whitman's homely symbol of unity); that plants and fungi, once thought to be at cross-purposes, communicate and collude for survival. Maybe it's radical to lend a tree agency, or microscopic fungi; more urgent to know intuitively this is so. And perhaps more difficult, since we witness the effects of a system of signals, not the signals themselves, as if one cried over a story one never heard. The white fibers of fungi are so delicate, they break when someone prods to investigate, fragility rather than brute strength safeguarding their ancient language.

As droughts worsen and temperatures rise and shorelines erode, some species of trees move north, invading areas where existing trees struggle to acclimatize to change. Some will fail, others will drown. Perhaps the void, which is not yet truly a void, will be filled. Weakening trees and dead trees may pass on their energy and emit underground alerts about insect attacks and pollutants and salinity, triggering resistance in new trees trying to take hold.

Millions of people in equatorial countries will also face droughts and famine and migrate north. I picture, not a global safety net, not a network of aid unbounded by nationalism and self-interest, not the altruism of trees, but the Gulf Stream as it clashes with the Labrador Current and glacial meltwater from Greenland, wars of incalculable chaos.

———

When wind whips up and rain pours down, solid as a sheet of tinfoil, I don't know what creatures take cover in the dying white ash or under its fragile ledge. I don't know which species are full-time residents though I imagine chipmunks. Garter snakes, which I sometimes find flattened on the road nearby. Salamanders and tree frogs. Toads. Woodpeckers are still to come when the sapwood atrophies. Or an owl.

If the tree were in California, its value as a home for wildlife might not match up with its risk as tinder. As the numbers increase, so do our fears; a mountainside of dead trees seems a preamble to destruction, already a stretch of tombstones, gray and scattered awry.

Since 2000 an average of 7 million acres have burned in the U.S. each year, more than doubling the average acreage of 3.3 million in the 1990s. The number of fires has actually declined, but their severity increased. In 2022 7.6 million acres burned, nearly half of that land in Alaska. In Australia bushfires ignited by droughts and temperatures of 116 degrees consumed 2.5 million acres in November 2019. Koalas tried to escape by climbing high in the canopy and waiting, curled into balls, but intense heat killed half the population around Port Macquarie north of Sydney. After such an inferno, what then? What to do with a snag forest that jars the eye and lingers year after year?

Clear-cut before charred trees and debris ignite again! Advocates of this approach say logging churns up the soil, which may be hardened after an intense fire, allowing new plants to root; they say the clutter left from cutting diminishes erosion. After massive fires they recommend that humans jump in and plant new trees, but one fire ecologist told *Yale Environment 360* that "…clear cutting and spraying herbicides amounts to kicking a forest when it's down." Logging destroys new seedlings that would serve as seed crop, and stripping an area affords greater potential for erosion and runoff. Herbicides attack native species and invite invasive ones. With time, pine and birch and oak will regenerate naturally on their own, and all the while snags offer homes for woodpeckers, fire beetles, owls, flying squirrels, ants, salamanders, and more. Areas with newly planted trees neatly placed in rows offer no such refuge, no diversity, which is vital for the survival of these creatures, often dispossessed by our actions in the first place. But we are short on time.

Snag forests comprise a multi-layered history: leaf litter on the surface, where salamanders may quietly control the species that consume it, and beneath it, soil teeming with bacteria, mites, fungi, and arthropods, soil in part fueled by the decomposition of dead wood. Clear-cutting erases the legacies that one generation of a forest might pass to the next. It reminds me of the furor with which my mother, after my father's death, stripped the house of his suits and socks, boar's bristle hairbrush, handkerchiefs, loafers and sneakers—all tumbled into Glad bags, firmly knotted, given or thrown away. When she died twenty years later, I went more slowly, sometimes saving things for which I had no use but imagined maybe my children would. Lost, though, were answers to questions I wished I had asked, unrecorded years and stories, thought shards, detritus.

I doubt anyone will notice when my dying tree dies on its cusp of shoreline. It may outlive me, may feed others after I can be of any use to my children. Maybe there has always been this ragged imbalance on the margin where survival meets extinction, but the process was slow and for eons unrecorded. Given time, given millions of years, a white ash might become as tolerant to salt as a loblolly pine, a beech might learn to live on a fraction of the water it now consumes. But we've accelerated the pace so far beyond the ability of living things to evolve that they, like humans facing droughts, are on the move, faster than ever before. Families at our borders are being torn apart; so too are forest species ecologically linked for thousands of years. From old-growth forests of the northeast, hemlocks are edging north, while beech are quietly heading west. As water bites into the woods where the white ash stands, and the forest recedes, phragmites will invade, emitting chemicals that kill off other plants. They will also protect the shoreline from rising water and catch the sun at the end of the day, setting the wetlands ablaze with light.

Facetime with a Mole

"LINA, LINA, COME QUICKLY." Ella tore across the grass and grabbed my hand. "Come see what's in the pool filter. There's a humongous spider and …."

I knelt on the hot stone, opened a lid, and peered through rippling water that ushers stray leaves into a skimmer basket. A spider nearly the size of Ella's palm perched on its rim, quietly alarmed or placid I didn't know. Beneath it, wavering among sodden pine needles and torn leaves, was a puffy dark body, not scrambling, not swimming. "A salamander," she stated. "It's dead."

When I reached for a trowel and dipped it toward the spider, Ella jumped back, shoulders shrinking as she clutched her arms to her chest. "The spider is fine," I said.

"But it's huge," she exclaimed.

"Yep, it needs to go back in the woods."

As I tried to nudge it onto the trowel, I held my breath, truly hoping it wouldn't dash up the handle or exit onto the stone at our feet. I had taught her spiders help us by catching mosquitoes and flies; we had marveled at a web spun between a deck post and coiled hose, dappled with water droplets like tiny orbs of mother of pearl. I could not display trepidation or even mild distaste. The spider sidled onto the trowel head and stopped as I lifted it and walked swiftly to clumps of wild blueberry bushes bordering the grass. When I turned my back, Ella dipped her hand in the basket (a hand still so small that knuckles indent the backs rather than rising like so many peaks), lifted out the salamander, and cradled it in her palm. It shone dully in the sun, its spots muted—black and gray spots faded by chlorine, maybe.

"Look, Lina, look! It's so cute." She tilted her head and raised her palm closer to her face. "Why does it have a long tail?"

"If a bird goes after it, it can lose its tail and still live and slowly grow another," I remarked, realizing I hadn't answered her question.

She didn't take her eyes from her hand. With the other, she stroked the belly. "Do you want to touch it?"

"Sure," though in fact I didn't. I ran a finger along the body. "My brother and I used to catch green and red salamanders in the lake I told you about. We'd make a habitat for them in a baking dish with some moss and lake water and keep them for a day or so."

"And then what did you do with them?"

"Put them back."

"Why?"

"Because they need the lake water, with stuff we can't see, to eat and live. And when they're bigger they'll want crickets and slugs and worms that we'd have to find."

Those salamanders were sleek with taut little bodies and a random arrangement of dark spots down the back. As I touched this one, dead for a day or two, I shuddered. The pale underside was soft and full, too soft.

"Can we keep it?" Ella asked, her hazel eyes wide, slanting sunlight catching flecks of green.

"Um, not inside, and you should wash your hands." Which is not what I wanted to say.

We opened a weathered door to an outdoor shower, and Ella placed the salamander serenely under the shade of a rhododendron bush. We rinsed off, and then she picked it up again, counting its toes, taking in its closed eyes just slightly raised from the head, the motionless legs, belly—bloated, I thought.

"Lina, why don't you like it?"

"I do," I insisted. "It's super interesting, it's amazing. I'm glad you can see it." Her skepticism was obvious, her intuitiveness on point.

"We can give it a little ceremony." (I saw myself at her age putting dead turtles with their backs gone soft into a shoe box and burying them in the garden and uttering prayers I didn't believe). But this idea would be meaningless to her. I turned off the water and watched her still examining the now warm body.

"That kind of salamander doesn't have lungs." She looked up, startled. "It sort of breathes through the skin of its mouth and neck, which have to stay moist. If the salamander dries out, the skin can't take in the oxygen so well. That's why they're usually out at night."

The next morning Ella ran to the shower for another look. After three days, even in the shade, the body turned ashy. "Doesn't smell too good," she noted. Without another word, Ella threw it over the rhododendrons and into the woods.

That night a thunderstorm hit around 2 a.m. Massive cracks of thunder, blistering lightning. Ella slept. I leapt up as I heard rain cascade through the screen door and beat the floor. The storm didn't cool the days that followed. Nor did the humidity relent. Temperatures hung in the 90s, the air was swampy, the woods didn't dry, and there in leaf litter burrows, salamander throats were pulsing.

———

In the following days, Ella checked the filters before cannonballing into the pool. A series of dead voles appeared, soggy round bodies with tight shut eyes nestled against the rim of the basket. One day a mom and a baby. Ella clambered to see as I lifted them from the strainer. She didn't hold them or wish to keep them. But she looked intently, very close, and I waited.

"Does it make you sad?" I asked, wishing I could see precisely what she saw and feel what she thought.

"A teensy weensy," she said.

"But you like to see what the voles look like?"

She nodded.

I couldn't let it go. Somehow all that death must be disturbing. I began sneaking out early in the morning to inspect the filters while she was eating breakfast or getting dressed.

The death toll mounted in ways I'd never witnessed. Three more voles. Seven baby tree frogs, their little toes frozen in place, their eyes still popped open on the top of their heads, their pale backs dotted, hind legs in place as though ready to leap. One morning a round furry ball pulsed against the side of the basket. At first I was unsure, my stomach suddenly stony. I went to get the trowel and gently lifted. So light. So soft I thought it wouldn't hold. A tug of water unfurled a wing, scalloped and finely webbed, wings that may have veered overhead the evening before as orange burnished the tree trunks and oak branches darkened. Ella might have seen this bat, that evening, or now, and I wondered how she would have reacted, what I was denying her or protecting her from.

═══

I've only seen one dead human, one I never knew in his lifetime. This was at a funeral of the grandparent of a friend. The corpse just looked to me like an old man with makeup. It left little impression. The friend stood there a long time, tears streaming as he bent down and kissed the cheek repeatedly. I wouldn't have kissed my grandparents, I thought, perhaps not my parents either though both were cremated. My children, yes, but this I cannot let myself imagine. I have no foundation for these biases, no substantiated reasons for not cradling the salamander body or calling Ella to see the bat. The fact that the pool this summer has become a death machine as never before should normalize our discovery of bodies, just as dying at home was once, centuries ago, a normal event that children witnessed—deaths of elderly, deaths of babies. Did that make the inevitable end any less terrifying, any easier to accept? But I get ahead of myself. Ella at five will likely not connect a vole's death with her own, and she definitely knows already that plants and animals die, so am I only protecting her from sorrow—not fear—and concealing, like so many layers of sedimentary rock, natural phenomena in which she is naturally interested, in fact, largely intrigued because of the ways we've explored woods and beaches together?

In high school biology, I refused to dissect a frog as assigned, the gangly white body floating in formaldehyde, the internal organs soft and useless as that salamander's belly. I refused not out of squeamishness, but in protest over raising animals for such purposes, protest against the industrial machine, which was profiting from that frog's helpless life. Now, however absurd, I felt implicated in unthinking destruction by creating a body of water with fatal chlorine and sheer gunite walls from which escape was as impossible.

The persistence of untimely death became eerie, in part because it was inexplicable. Storms battered us at night, mornings wrapped us in water-laden air. Surely the wildlife, seen and unseen, had water.

———

One afternoon while Ella was out, I stood on the top step of the pool, about to plunge in, when out of the corner of my eye I glimpsed a gray body squiggling fast and hard through the shallow end, heading along the wall toward me. Quick, the trowel. When the creature arrived at the top step, still wiggling hard, I tucked the trowel under its belly, but twice it evaded me. At last I had it and quickly jiggled it onto the pavement where it seemed totally disoriented and headed blindly back toward the pool, so I edged it again onto the shovel and flung it not terribly gently into the grass. Its head seemed stunted, eyes hidden by hair, white claws on paddle feet most prominent, wispy little tail. Within seconds it had concealed its body with grass. I bent to watch. A patch of clover and ragged grass rose and fell, rose and fell, as if the earth were breathing. It felt intimate, this small piece of earth rising and falling, whether because the mole was gasping from its swim or digging in rhythmic strokes, I don't know and perhaps there is little distinction in meaning, both being the continuation of that life. On it would go, making its solitary way into tunnels where it would eat worms and avoid other moles, dig and eat, tunnel and eat. Ours had been a close encounter. And I began to regret that I hadn't held it longer, noted its sixth claw, a sort of extra thumb that grows later, after the other five fingers. It develops from a bone in the wrist and acts as a platform while the mole digs. This is a type of poly-dactyly, a playful, rhythmic word that sounds more like an adverb than a noun and connotes a friendly dinosaur or the way one might play a dance on the piano. It can occur in humans but is often corrected with surgery. For moles, it's the norm.

I resolved to tell Ella about the mole and began to wish she'd been there. Clearly she had felt some mix of pride, validation, joy, and awe the day she rescued a dragonfly from the pool—so carefully extending her hand and cupping it in her palm. It didn't fly off as I expected but alighted on her arm for some time while she admired its iridescent wings with their intricate webs and shadowy tips, its needle body and bead eyes. "There it goes," she said at last.

More recently, Ella has been rabid to catch something live to keep, which might be a nurturing instinct or an extension of general acquisitiveness already seen with bracelets and stuffies. First she trapped a sand flea, a tawny half-inch crustacean, that crawled and hopped. As she went to place it in a bucket, it found a crack in her cupped palms and leapt out. Her expression, incredulous, fell as she moaned with grief. That same morning, the tide out and bay shore strewn with scallop shells, split razor shells, empty whelks, clumps of dead man's fingers fixed to rocks, and barrel weed undulating in the water, she found a large open clam shell with, oddly, a minnow inside. A treasure. I dashed twenty-five yards down the beach to get her bucket and fill it before the minnow expired and she followed fast, but before I knew what she was doing, she was at the water's edge, gently rinsing the sand off the minnow (which I found endearing if naïve) and naturally, just as I was filling the bucket, it vanished. Tears and more tears. She wanted to search for *that minnow*. She was sure she would never ever catch *another minnow*. It had been a fluke to find it. It had felt fortuitous, something I would never have thought possible. She went on checking other clam shells with serious expectation, just as she feels it's very possible she'll find a pearl in one of the beat-up oyster shells we come across.

We found minnows, but though I swooped with my bucket, they were too quick for me as I knew they would be. And then Ella, who always says she has her eyes on the ground while mine are with the birds and the sky, spotted the dappled claw of a fiddler crab partway out of its hole. With our plastic shovel, we lifted him into the bucket, along with some damp sand. And a little dry sand. We weren't sure of the proper proportions. And we set off back down the beach and over the dunes while the fiddler tried to scramble up the side of the pail, spreading its skinny black legs and looking like an arachnid, which it isn't. It wasn't a very large pail, and I wasn't sure how nimble the fiddler was, so Ella shook the pail from time to time to knock it off the side and back to the sand. This didn't seem kind, so we decided to go home and get a large bucket to ensure his captivity and also supply "Fiddie" with more sand and bay water, which I hoped would offer the microscopic things he ate. I scrubbed out a cleaning bucket and we installed him in the kitchen where for that evening Ella watched him and introduced him to her younger sister. Ella touched his claw and moved him around, clearly not worried about being pinched. I told her that we could keep him for a day or two and then return him, whereupon my husband who hadn't seen him yet but went online said you could keep a fiddler crab for three years. "Three years?" asked Ella, grinning at me.

After three days, I said, "I hope he's okay." She nodded. I wanted to return him to the water before he died, which I thought he would. But she didn't seem concerned. She seemed careless not about him but about death.

It was fiercely hot that afternoon. "It's time," I said, but Ella groaned. "Lina, could you take him back, *please*, it's soooo hot." She didn't want to walk the quarter mile down the road and then all the way down the beach to the stream where he lived. I thought she should see it through, take responsibility and all that, an adult construct that perhaps was irrelevant. She already thought he was doing fine. What did I think she would learn? She had studied his finely pointed white claw and observed him scuttle across the bucket's damp sand just as she'd seen scores of others do by the stream. Were he dead on arrival, should she know? Would I tell her? Would guilt weigh on her like Fiddie's ungainly claw? Sweating, I lugged the bucket, discarding some sand along the way. The tide was out, the sun ricocheting across glassy water. Fiddie had dug a hole in the damp sand and stayed tucked in there out of sight, if weakened I couldn't tell. When I reached the stream, I tipped the bucket and spooned out some sand till I found him, and he found the water, came to life and scuttled in, instantly flicking his smaller claw toward his mouth, as if ravenously hungry and grabbing anything within reach. He did this for a few minutes before crossing the shallow stream and hiding in the shade of some tall grass at which point I felt I could leave. Ella didn't ask what had happened (she was carrying around a tiny dead grasshopper), but I told her anyway.

———

I also told her about the mole. "Why didn't you keep it?" she exclaimed. "Oh, I wish I'd seen it. Was it fast?"

"Yeah. It was pretty rattled. And you weren't here."

"You should've called me!"

"And face timed with the mole?"

"Yes!"

In truth, I could have. I could have really looked at those near-blind eyes and scrappy extra thumbs as I'm sure I won't see another common mole on whose tunnels I walk every day, my foot feeling the earth exhale.

———

Late summer brought days so still the leaves never flinched, the sky lay pale blue behind a diaphanous veil of clouds. Ella and I spent hours in diamonds of turquoise light underwater, torquing like otters, pushing off from the bottom, diving for acorns, and holding hands while we said words to each other that neither could understand. "This is how I play," she said, spinning in the water and entwining her hands overhead, "while you do laps." During those days the filters caught only leaves. Catbirds and blue jays awoke casually, lacking the fervid eruption of songs in June. Cicadas chirring all day often went unnoticed. Russet hydrangea blooms lingered.

My daughter packed up the car and her girls bounced around in excitement at heading home to Brooklyn. Florence at two didn't know how long she'd been gone or when she'd be back. At five, Ella knew. She leaned against my body and looked up, "It's bittersweet."

Maybe that's one word for her thoughts as we discovered frogs and voles and fiddlers to observe and to live with and say good-bye to, alive or dead, though as she grew the inherent balance in the word would change.

Fin

AT 11 A.M. ON New Year's Eve my dog stares at me, square in the eyes, from under a brow of black curls as if he wants to know. He wears a translucent lampshade on his head to keep him from gnawing the stitches in his left leg where a mast cell tumor was removed six days before. During the night he scratched so hard he pulled off the collar and his ferocious licking of the wound woke my husband. He looks sheepish and slightly absurd, those perpetually gentle eyes with cusps along the bottom rim, milky and red, as he looks up. The area above his hock is spongy, the stitches bristling, the incision bright like coils on an electric stove. He itches all over, a stealthy legacy of the tumor, said the vet, and I give him Benadryl pills wrapped in bologna. After scratching to go out, he stands under the pine tree staring blankly, severed from his olfactory reality by a good ten inches. He scrounges through the ferns like a blind man in a supermarket. At night I hear his leg drumming against plastic in a fruitless attack. I hear plastic whapping into coffee tables and chairs and brushing and thumping the carpet on the stairs as the rim of the collar catches. Last night he barked at one, three, and four a.m., a piercing bark that swerved into my consciousness like a driver out of control. I tottered down the stairs and slumped at the kitchen table, blinking my sticky eyes, while he hustled outside. Through the black windows I couldn't discern his black form at all, only the ghostly lampshade moving back and forth across the yard.

At noon we are back at the vet's. I sit on the edge of a bench in the miniscule waiting room. Finny quivers and leans into my knees while the vet's pug scuttles around breathing heavily. It's a small-town family operation, the vet in jeans and his wife in a long flowy skirt with crinkly gray hippie hair down to her waist. With her hearty arms, each bearing at least a dozen tin bracelets, she lifts Finny onto the table. Somehow the incision burst open and wet flesh like a persimmon glistens under fluorescent light. Fuck, says the vet. You'll have to leave him here for a few hours. The malignant cells started as immune cells to combat an allergy, he explains. Now they've shifted allegiance and traveled to the perimeter of the battleground, possibly launching into new territory. The vet sighs and writes prescriptions for tranquilizers and antibiotics and recommends we start chemotherapy at the end of the week. Since the treatment is systemic, I decide against the blood test that would reveal if the cancer has spread.

Nine years prior my father was diagnosed with inoperable lung cancer. We expected a long, slow, painful decline. We expected to wait helplessly at the mercy of aberrant cells. Within three months they traveled to his once acute brain and killed him. But we learned that later; I don't think he ever knew.

By 4:00 p.m. Brian has brought Finny home, and he lies at my feet. "Fin, Feeee-un," I say, stroking his flank. Lately he has sensed a new poignancy in my touch and capitalized on it, leaning against my knees in a near swoon. But now, basking in a bath of refracted light and still drugged, he doesn't budge. The light reveals my fingerprints on his collar, flecks of dirt, and dried streams where water dribbled from his chin. But he's only six.

Around 7 p.m. Finny heaves himself up and sways to the top of the stairs. His front paws extend in front of him, and he dips his head in a futile attempt to lick the shaved area where the vet inserted needles during surgery. Poised like a sphinx, he stares down the stairs toward the front door, his eyes half hidden under a curtain of black curls, his white chest bright like a dress shirt, anticipating something—it's New Year's Eve after all—and we're all balanced on the fulcrum between what we've done and what we will do.

I look back, Brian ahead. I picture Fin outside when the kids were little, his forelegs scrambling rapid fire as he dug up daffodil bulbs and left their brittle skins among the daisies, or rolling in grass seed, leaving craters of naked dirt. He scaled chicken wire and dug under a post and rail, ate a rope hammock, butter off the counter. Now Brian is staring into the screen of his laptop as if into a crystal ball, reading about tumors, grasping at a prognosis. His eyes water and he claims he forgot his allergy pills. "We don't *know* everything yet," I say.

"It is what it is," he shrugs, not quite believing that.

Then he eyes my plate because I never eat the last bite and he knows who will. Since it's New Year's Eve, there will be steak scraps for Fin—fat pure and simple— unencumbered by time, an assessment of accomplishments, or resolutions for tomorrow.

At 11 p.m. the phone rings. It's my brother in Wisconsin who wishes me happy New Year, prematurely, and tells me he has prostate cancer, the most aggressive kind. If it hasn't spread, they can do radiation; if it has, there isn't anything they can do except, says the doctor, pray, the tacit understanding between me and my brother being, no one's doing that. It would be another waiting game, an endgame crazily out of our control. Pacing with my phone, eyes blurring like snow outside a dark window, I almost trip over Finny despite his lampshade collar. He has rolled on his back in anticipation of a belly rub and eyes me with near disdain—you can't deny me. Silence on the line. I don't know what to say. My brother. Except I can't believe it. Survive. Don't be anxious about what hasn't happened yet. I squat and rub the slope of Fin's belly from the smudgy white chest to the thinning hair near his groin.

The confluence of events is fictive. Who would believe it?

New Year's Eve is long. What's quick is the flick of the red clock at midnight, then a hiccup in time before we ask, where are we now?

On TV crowds are dispersing from Times Square, and a blizzard of ticker tape lies underfoot. White blood cells and red, bone cells and marrow, obedient cells and cells in riot are preparing to divide.

═

Labor Day. Fin has dug a nice cool hole in the dirt against the wall at the back of our property. He peers at me from behind the trunk of a pine that has lost its lower branches. He watches my every move now, following me when I go inside, standing at the low window and staring out when I get in my car. I don't know what he needs now that he didn't before.

The texture of his hair has changed. It's lighter, less curly, less black though not gray. I try Black Pearl Shampoo and Conditioner but it doesn't have lasting effect. Mulch from the yard and leaves that fell early because of a drought cling to his coat and litter the house when he comes inside.

It's been five days since the last surgery, and the lampshade lies in the corner of the kitchen. The vet said we wouldn't need it at all this time, that Fin was so old and stiff he probably couldn't reach the stitches near his anus where another tumor was removed. But as soon as the anesthesia wore off, he was at it, curled in a ball, gnawing at the stitches and licking with all his healing power focused in the muscles of his tongue. He's a survivor—a scraggly fifty-pound dog who has undergone four surgeries and a month of chemo during which he lay splayed on the kitchen floor, failing to respond when I'd put on a leash and give him a tug.

As the sun rages and temperatures accelerate to three digits, there he lies in the dirt, deaf to my calls, deaf to convertibles shifting to high gear, music blaring. He has retired from long walks up and down hills and into the woods. The circumference of his world is shrinking, like my mother's.

Last night she went to the hospital again. She doesn't swallow well, and some bit of bread or the tip of a bean lodged in her lungs like a fly on flypaper whose wings vibrate uselessly as in a dream where the will cannot will one's feet to move. And so the lovely bit of white bread caused vomiting and a fever of 104. She was so weak a nurse at the nursing home called to ask my permission to send her to the hospital. According to her living will, which she wrote years ago when her mind was as sharp as an eagle's beak, she doesn't want feeding tubes and resuscitators; more recently, in consultation with her doctor, I signed a document about "comfort care" which discourages undue intervention that would be stressful for a woman of ninety-four. So now they have to ask my permission to save her life.

I wouldn't have let the vet give chemo to Fin if it did to dogs what it does to humans. At some point one suffers enough. And a dog suffers without knowing why, which makes it worse, some humans think, though a dog doesn't need to contemplate the odds of survival. Still, I look at him and think, I'm playing god: I decide you should live longer, or try. Or I presume: you can't enjoy this life, you must be in pain when you get up from a nap and your hind legs buckle, I will end it tomorrow. But maybe the humans aren't ready, aren't quite sure.

The first time my mother was in the hospital with aspirated food causing pneumonia, a doctor called me and asked about ventilators. I was driving to work. The road curved under overhanging maples that blocked the speed limit. A cop passed me, glaring. I didn't know about the living will or if I did I couldn't think of it *then*. Should I say yes to breathing? I pulled off the road, put my head on the steering wheel for a moment. "Yes." But then I came home and dug a yellowed document from the back of a file cabinet and called him back.

Even so, she survived.

By now dementia has thwarted any attempt to know what travels through her mind as she dozes in a wheelchair before a neon-blue fish tank. I don't know if she fears death, don't know if her picture of what awaits at that end of life has changed. Maybe she clings to glimpses of the living world she still loves—my brother, me, a bed of white iris out her window. Maybe she remembers angrily her vehement demand years ago—never let me end up a vegetable! She can't tell me if the rules of the game have changed; my brother and I made the uneasy decision not to hospitalize again, not to operate again. At some point one suffers enough. Last year at the nursing home, two sisters—twins at the age of ninety-five—hoarded pills and attempted suicide. A friend betrayed them to the authorities, who intervened, separated these dignified, fully lucid women, and whisked them away to two hospitals where one survived and the other did not. They told the clientele at the home a different story.

Any day Fin may stop eating. Some animals do. We won't know if he has a premonition of death and wants to die, or if he feels so sick he just doesn't want to eat, and will, then, die.

Since the result is the same, it should be easy to brush off the distinction, but I can't.

My brother finds himself wishing he *only* has prostate cancer. My mother drifts in the midnight zone, surfacing to nod at an aide or snap at a woman sobbing at her table. Fin lies on the cool stone floor, slapping his tail now and then. Sometimes he pushes open the screen door to dart after a squirrel for a few yards, seeming in his present-ness to own his future more than they.

———

By 7:00 p.m. oak branches are swirling and gray rain coats the windows. I go upstairs to get a long-sleeve shirt, and as I push aside my summer dresses, I look down and there is Fin. He has found me, and I give him a quick pat on the head. He looks up from under his black curls. One eye is filmy blue like the windowpanes, and white pus lines the lower rim. With a pat he is reassured of whatever it is he needs now and didn't need before. Given the untraveled realms of his consciousness, I don't know what that is though in my own mind it borders on trust that I can do *something*, a reassurance I couldn't offer to my father, nor now to my brother or mother.

Out the window I see the birch has already lost so many leaves, they lie around the trifurcated trunk like paper tears. One trunk stands half the height of the others, amputated in a freak snowstorm one year when the leaves were still on the trees and snow weighted the limbs till they snapped like chicken bones. Because fall is coming, we'll hustle to the train tomorrow with renewed vigor. Because my mother is ever so slowly dying, I'll call tomorrow and wake her up. Because Fin is incontinent, we'll roll up the rugs in every room. We sit down to eat as lightning flares through the house, lighting for an instant the green glasses filled with water, the softening butter, the burnished kernels of corn, the fat-veined steak. Fin takes his customary position at Brian's side, pants for a few minutes then falls asleep, certain of certain scraps.

Like a Sycamore, Like a Laurel, Like a Dove

THE SKY WAS FLAT as gray construction paper, a film of snow lay on the road, sleet hissed and pricked at the windows. It had been cold for two months, and I had no hope of warmth until May. So when a surprisingly clear morning in February arrived, I set out to walk in some nearby woods. Around me were tangled vines, naked oaks, sorry maples, all so brown and monotone, the world seemed reduced to a single dimension, a *Waiting for Godot* sort of stage. Then as I reached a clearing, I saw above the morass of twigs, sudden as the opening chords of a symphony, the splayed white branches of two sycamores backlit by the sun in a rich blue sky. They shone, swayed slightly and silently. They demanded my eyes, they were sleek and clean as bone, but not skeletal. Had this been August, all branches laden with tired leaves, I wouldn't have noticed the trees; they would have blended and past I would have trudged, wiping the sweat off my forehead. Had the day been overcast, their effect might have been commonplace, less uplifting. Instead, the trees reminded me how

often I've admired sycamores—at various twists and turns in New York's
Central Park, along countless Main Streets, USA, on hilltops and river
valleys, sporting their dangly fruit, stretching their oddly curling branches.

My first error was in believing sycamores to be virtually ubiquitous.
The American sycamore, native to this country, likes rich alluvial soil
found along streams and floodplains. They're also found in graveyards
because, according to Luke 19, a tax collector named Zacchaeus climbed
a sycamore to see Jesus as he walked through Jericho and was greeted
warmly. Thereafter, the long-lived trees came to symbolize a link to the
divine for the dead who await a Second Coming. In fact, the original was
a sycamore *fig*, which is native to the Middle East and parts of Africa.

Furthermore, all those trees lining the sidewalks of Manhattan and filling the squares of London are not of the floodplain variety, the American
strong man, largest tree indigenous to eastern North America. It's probably
presumptuous to think so. Rather, the sycamores that withstood the choking soot of industrial London, survive cramped cutouts in asphalt in cities
across America, tolerate massive clipping to dodge phone wires—these are
London plane trees, a hybrid of the oriental plane tree and the American
sycamore. The breeding allegedly occurred in the gardens of Charles 1 at
the hands of his gardener James Tradescant the Younger, who visited the
colonies in 1636 and brought back some saplings; there, the American type
mixed with the oriental variety, which had already immigrated sometime
before. When the trees had matured enough to bear seeds, the London
plane traveled across the Atlantic to the land of half its genetic history
to become a staple in American cities and parks and to be confused with
its native relative, a controversial sort of assimilation. Whenever I saw
exfoliating bark, that army camouflage look, those gradually exposed
limbs, I thought sycamore, but now I know better: the lower portions of
sycamores have uniform bark while the London plane exfoliates from its
top to the base of the trunk.

Even if I'd noted the difference, I might have thought the disparity nothing more than an arbitrary choice of dress, that is, wool skirt versus shredded jeans, from which metaphors begin to unfurl. What were the trees reaching for? Or what were they celebrating with their long arms, their fine fingers? I stopped that day, looked at the uprush of branches, and thought about Wallace Stevens who claimed, "in the presence of extraordinary actuality, consciousness takes the place of imagination." The exalted streaks of white rising above everything in my line of vision *were* extraordinary, and they were branches, nothing more. Yet I persisted (my second error) in turning the sky into construction paper and dressing a sycamore in army fatigues—and failed to question if such images arose from facile indulgence or whimsy or a deeper desire for affinity. As Emerson wrote with characteristic certainty, "Nature always wears the colors of the spirit."

Humans draw analogies between themselves and trees for obvious reasons—trees being rooted, individualistic, potentially strong and long-lived—and the transfer of meaning can run in either direction. Robert Frost didn't like the idea that ice storms bend birches, though they do; he preferred to think about a boy, himself, swinging from the trees and slowly subduing them. And once bent from years of such play, the trees drape their leaves across the ground, just as girls "on hands and knees" fling their hair over their heads to dry in the sun. As a nature-loving adolescent, I always liked "Birches" and heard fragments in my head as I walked through coniferous woods, somehow missing the slightly sinister intimations of power, sex, and dominion.

In Ovid's tale of Apollo and Daphne, the god's vision intrudes on his object of passion with images that belie the gritty reality of the situation. As Daphne, who wants to remain a virgin and has no use for men, flees the ardor of Apollo he calls out, "Nymph, Wait! This is the way a sheep runs from the wolf, a deer from the mountain lion, and a dove with fluttering wings flies from the eagle: everything flies from its foes, but it is love that is driving me to follow you!" Like a sheep, like a deer, like a dove! Apollo's analogies are wildly disparate, taking little account of Daphne's character or her plight. Still more absurd and blatantly ironic, he claims that she is *not* prey, that he is *not* a predator, when just a few lines later Ovid writes, "Like a hound of Gaul starting a hare in an empty field, that heads for its prey, she for safety: he, seeming about to clutch her, thinks now, or now, he has her fast, grazing her heels with his outstretched jaws…" Exhausted and overpowered, Daphne pleads with her father, a river god, to transform her into a laurel tree. Her swift feet become imprisoned in the earth, her face hidden, her limbs numb. And yet, Apollo thinks (as he describes the honorable future he envisions for her) that she "seemed to shake her leafy crown like a head giving consent."

A more cryptic tree analogy describes Hester Prynne after she suffers for seven years, ostracized and shamed with an "A" blazing on her chest: "All the light and graceful foliage of her character had been withered up by this red-hot brand, and had long ago fallen away, leaving a bare and harsh outline…." The compact, concrete image makes us leap to the idea that she has lost her physical beauty while the words say the beauty of her *character* has withered due to the Puritans' penalty. So which is it? Hester has in fact lost her physical beauty—her face ashen, glossy hair tied up—while in isolation her thoughts have blossomed beyond the scope of imagined tree limbs, beyond the borders of the Puritan town, to take issue with the entire hierarchy of male/female power. She is wiser and more confident than ever. Hawthorne's male narrator conflates character and physical beauty in our mind's eye, leaving us a forlorn remnant, "a bare and harsh outline," (the maple in winter having little to offer). He skirts an outright embrace of the heroine's intellect with a reductive metaphor that betrays her.

———

I used to think, as a teenage pseudo-Romantic, that my singular self was expanded rather than reduced by moody associations with ragged mountains or diamond patterns of bark or winter-swept trees scratching at the moon. I wrote poetry that purported to translate the world into words and (more crucial to any adolescent) to define myself in terms of that world, make it more palpable while remaining unknown. As if some bulb of essential being lay hidden for me to unearth. This activity heightened my appreciation of the unwieldy variety of elements under the umbrella term, *Nature*. It made the wilderness where I hiked somewhat more familiar, a probable impetus for anthropomorphism. And it helped me believe I *understood* trees and flowers unknown to me, as Apollo blindly thought he did in the presence of a laurel. Or any man does who fashions tree women.

I felt an almost visceral exchange as I walked through a forest and thought about threads of human experience: that dead limbs stand like a crucifix (the leadwood of South Africa die and never fall), that old limbs jut like elbows, that saplings scramble for light like siblings for attention, that an oak endures, its arms repeating cycles of birth and loss over and over again, decades after I decide its branches are like cradles, its roots like memory—though this dialogue was clearly one-sided.

At eighteen as I hitchhiked around Alaska, scribbling poems and snapping pictures in an attempt to capture something of that massive landscape, I had what seemed to me then a revelatory thought:

The clouds fold and touch as hands
barely

and smooth
the sloping shoulder
green as woman
and foothill sinews
wind-hewn man

if man weren't man
the moon would have no face
hills would be hills
and cloud would be clouds

The poem itself isn't noteworthy, but back then its directness surprised me, the second stanza so unlike my usual tangled abstractions and thorny imagery. This one felt cathartic, at once obvious and profound, freeing not only me from a specious search for self-definition, but also the land and sky, releasing it from cumbrance and clutter—like stripping off a winter coat on a suddenly hot day. Or like a sycamore shedding its bark and exposing its vital inner bark to oxygen and sun. Yet we don't know definitively why it does. Today when I see sycamore branches against the sky or pass a plane tree on a city block, such uncertainty, even mystery, enhances the tree's autonomy and further exposes the presumption of seeing ourselves in its branches and bark. Or the moon as a man.

Although I recognize the potential of metaphor to elude or deny, the impulse to connect remains wired in my brain. Always an urge for more, though more of what is uncertain—if not meaning (which was tossed like a gum wrapper from the window of a car by postmodernists decades ago), then possibility. An open highway. A night when a full moon emerging from behind a cloud lends a third dimension to silhouetted pines. A new vantage point. Metaphor inhabits a space in the mind's eye where we light upon a connection and question what kind.

In 1971 some sycamore seeds took a ride on *Apollo 14*, which circled the moon. Part experiment, part publicity stunt, the Forest Service wanted to see what would happen to seeds exposed to radiation and zero gravity; a former employee, smokejumper Stuart Roosa, was an astronaut on this voyage. While two astronauts disembarked, walked on the arid surface, and hit a few golf balls, the seeds of five tree species stayed aboard. Back on earth, they were planted in areas across the country in environments ranging from Koch Girl Scout Camp in Cannelton, Indiana, to the Goddard Space Flight Center in Greenbelt, Maryland, to shopping center megalopolis King of Prussia, Pennsylvania. By and large, the trees flourished and gained notoriety as the "Moon Trees." The seeds had been aloft and returned. I think about the journey of sperm to egg, egg to uterus, and the descent. I haven't read that the seeds had better chances of surviving on earth after floating in space, nor can we measure what knowledge of sound or sight or spirit an infant brings from the womb. The analogy is a stretch, but it reminds me that metaphor is a journey taking one *almost* there, the body being the body, (the tree the tree), the thing to which it is analogous being departed, on the road, in space. In his poem "An Ordinary Evening in New Haven," Wallace Stevens wrote:

Who has divided the world, what entrepreneur?
No man. The self, the chrysalis of all men

Became divided in the leisure of blue day
And more, in branchings after day. One part
Held fast tenaciously in common earth

And one from central earth to central sky
And in moonlit extensions of them in the mind
Searched out such majesty as it could find.

The self is tree-like, branching in different and disparate directions—one part rooting deep in the earth, grounded, itself and nothing more; the other part living in a nebulous region between earth and sky, a dimly lit reach of the mind that finds, not "majesty" entire, but what it can, which suffices to keep one looking. Despite the difference, each mirrors the other in a sort of amped up version of embryonic division—roots forking underfoot while "moonlit extensions" of thought branch like neurons casting nets in space. I suppose the creative instigator is us, you or I or anyone, who has traveled in unpredicted directions on a random February day and witnessed, say, a bouquet of bone-white branches.

＝

And yet, despite the alluring breadth of earth and sky, flesh and spirit, Stevens's dichotomies leave the self divided, restless, and unsettled. Recently I came across a poem called "How Plum Flowers Embarrass a Garden," by Lin Bu, who lived during the Northern Song Dynasty roughly a thousand years ago. He so loved a plum tree, which burst into bloom in January, that when he retired, he lived alone by a lake where he wrote about its white blossoms, painted them, and ate plums until he died. According to legend, people would ask him, "Don't you want a wife?" to which he would respond, "The tree is my wife."

When everything has faded they alone shine forth,
encroaching on the charms of smaller gardens.
Their scattered shadows fall lightly on clear water,

their subtle scent pervades the moonlit dusk.
Snowbirds look again before they land,
butterflies would faint if they but knew.
Thankfully I can flirt in whispered verse,
I don't need a sounding board or winecup.

Without analogy or verisimilitude, Lin Bu unleashes our imaginations to beauty that startles the snowbirds and evades the butterflies. His words "…if they but knew" hint that he himself might, and yet the intimation is as humble and unintrusive as a high-flying cloud one scarcely notices on a summer day. Could any tree, so adored, object?

Ghost Plant

I HEAVED OPEN THE bulkhead attached to the north side of my house and descended five steps, hunching in the musty low-ceilinged cellar. Indirect light from the entrance caught something feathery and white. As my eyes adjusted, I found two stalks rising a foot or so from an old flower pot half filled with dirt. Each thin spear bore a moon-white flower, or memory of one, an eyeless sleeping head. I had no idea what had grown in the pot the previous summer and what could have generated in absolute dark without water. I didn't know if it was alive.

I lugged the pot up the steps, squinting at May sunlight that washed over the yard and turned the grass greener than it was. The stalks shivered in the wind, the blanched flower held fast. I dug out the threadbare roots, ashy dirt tumbling between my fingers, and carried it to the other side of the house where a haphazard garden meets the woods, planted it there, and went on with life.

A month later the spidery flowers had vanished. Thin green spears clustered at the top of the stalks while around them five new stems had sprung up, each with a balanced array of lateral leaves. A few skinny petals spiraled from the stem. I'd never seen a plant like it and no one I knew could identify it. One morning, as I crossed the grass that had turned the color of oatmeal toast, I ducked under a pine branch and stopped short. Standing by the ghost plant was a huge hawk who stared straight at me, instantly cognizant and territorial. It expected to stay and me to leave. We stared until I backed away, leaving it to listen for mice or read the paths of voles.

The plant reached about eighteen inches with five stems stretching from a single root system. Its spindly leaves were unremarkable, but it produced three tiny flowers, robin's egg blue. The following March I looked for it to nudge up through the well watered soil and feel glimpses of sunshine, as I was doing, but nothing appeared. A plant that had enacted such a bizarre resurrection in the destitute conditions of my cellar should thrive year after year. Such was my expectation.

That month, my daughter-in-law was seven months pregnant. The baby was kicking, growing, all in the realm of normal. Scans showed a face at rest, slit eyes and wide nose, strangely coated in gold. My husband said he looked like our son, but I couldn't see it. Suzanne was scheduled for a C-section in early June. Hearing that thirty percent of women giving birth at her hospital, Mt. Sinai in New York City, had COVID-19, she began to pursue other options. All expectations of delivering with someone she knew and trusted were now in limbo. I wondered how many women would not find a hospital bed when their baby, growing cell by cell safely in the dark, came due. Friends of my daughter said they wouldn't get pregnant now. I don't know if I would have if times were like this back then.

═══

April arrived—nearly a year after I found the ghost flower in my cellar. Although that winter, 2020, was the second warmest on record, spring malingered. Eighteen-mile-an-hour wind, cloudbanks solidifying the sky, oak branches bare, phragmites still khaki along the inlets. Only forsythia perfunctorily graced my neighbor's weathered fence. The earth would rotate, its axis tilt, snowdrop petals bow over dead leaves in the woods— these uneventful acts ground against human paralysis like slipped gears. We waited on our missed dreams and texted our sick friends.

My daughter's one-year-old daughter sat on the grass behind our house and discovered broken acorns and dead twigs. I couldn't look away for a microsecond or she'd pop a jagged cupule into her mouth. Acorn, I said; Cacor, she said. Sky, I said pointing up. Kye, she pointed. Every day we named things; every day I said what is this, looking at pictures of cows and dolphins and donkeys, or can you say this, raccoon and duck, though sometimes I just bantered in her language as if I understood all she meant to tell me about her day—doggie (accent heavily on the second syllable), doggie, doggie, daga, daga, ga, gaba. Dog dissolved into something I didn't know but she did and our conversation never flagged.

Rock, I told her, as we passed boulders along the shoreline and I stooped to let her feel the texture of feldspar and see flecks of quartz. All this naming was affirmative, as much for her as for me. For an hour we stitched the world back together. Name and repeat, name and repeat. I told her the birds fly, the ducks fish, but those actions didn't yet have meaning. Nothing needed to be doing. All needed to be existing. As we walked down the road passing stiff bare oaks and wafting white pines, she exclaimed dewee, dewee with exuberant emphasis on the second syllable, eee sailing up to the branches as she pointed. First thing in the morning she shouted dewee and pointed to the window. And I said, yes it's still there.

Sitting on the damp ground, still in our winter coats, I watched Ella pluck weeds and glide her hand over the grass and slowly finger an acorn in a world that was emotionally and morally stable, not indifferent, nor empathetic. Acorn, cacor, name and repeat, name and repeat. She liked to say happy, happy, though she didn't know what it meant. I wondered how to teach abstractions or if I wanted to. The scaly cupule in her hand protected the embryo in a kernel that may have grown elsewhere or not. My lightless cellar harbored a seed that had no reason to sprout. Viruses thriving in one place or another had no reason to flourish in millions of lungs and throats, and so the pandemic seemed not just tragic and terrifying, but absurd. Unapologetic survival is reason enough, I suppose, but the wild card is chance, the random threading of factors that determine whether a weed happens to bloom, a man happens to live.

A new planet sits over us, auburn and blurred in the sky. I hold my breath when a man passes me in the supermarket as I reach for a container of milk for Ella.

═══

During the first week of April 2020, makeshift morgues lined East 30th Street in Manhattan, and a fleet of refrigerated trailers was bound for the five boroughs. Forty-five mobile morgues would house 3,500 bodies without light or ceremony. While packages streamed along Amazon's conveyor belts, bodies flowed from hospitals to funeral homes to crematoriums to cemeteries. We wondered where a backlog might occur, and when the bodies would wait in the streets. At that time my inbox was studded with requests to join chain letters, a phenomenon I hadn't seen since high school. I was supposed to send a meaningful message or quote to the next person on the list and wait to see what would come to me. Neediness or thoughtfulness or boredom flourished in isolation; the sky was knitted with messages, platitudes, panic. Mapping of the virtual and physical landscape had changed. It was no longer someone you knew who knew someone who was sick. The universe was involved. Three of my friends' parents died within a week—from cancer, pneumonia, and stroke. Suzanne's baby was twisting and turning. At night she could nearly see a foot shoving from her abdomen. Daily she looked larger, scarcely able to tie her shoes. The baby crowded her other organs so that she couldn't eat much and felt nauseated when she did. Relentlessly, the baby grew, as he should. Miraculously, the baby grew, as they do.

On the day Ella was born in April 2019, *The New York Times* front page headline read, "Peering into Light's Graveyard: The First Image of a Black Hole." For the first time scientists were looking into a space so dense and unfathomably deep that it swallowed light for all time. But the photo seemed to me like a gaping pupil within a fiery donut rimmed in red that stared blankly at us. *Ha, I exist. I exist in a galaxy you choose to call Messier 87; I exist 55 million miles from you. I am several billion times more massive than your sun.* Beneath the photo was another, just as beautiful. Blue-white curves and sensuous folds on a night sky. At a glance it might have been clouds, ephemeral and insignificant. Or an effusion of smoke from an atomic bomb test in the desert in pictures we used to see. But no. The image was a glacier, backlit, glowing, silver blue—a glacier "Melting Sooner Than Expected," issuing isolated chunks of ice into the dark. Fire and water. Elegant and elemental.

Precisely one year later, April 11, 2020, a smattering of headlines hits the *Times'* front page: "Burning Cell Towers out of Baseless Fear They Spread the Virus," "Virus Deaths Mount…," "Global Trade Sputters, Leaving Too much Here, Too Little There," "Torn Over Reopening Economy…." All chaos, rupture, disunity, uncertainty, and urgency to cover what is too fractured to cover. We're scrambling, feeling at last as small as we are from the eye of a black hole that devours light, leaving us hoping that for no reason life in the dark (without water, air) will win out. In the nowness of her short life, as we hold our breath, Ella finds verbs: Eat... Try... Run.

PANDEMIC TIME OR
A CAR-LENGTH STORY

Just about anyone I talk to says the pandemic threw off their sense of time. Now in 2024 I find myself telling the guy who cuts my hair, yes, I had COVID, then a pause as I try to remember, was it January 2020 or '21? Maybe it was the scotch I drank that blurred those years, but I think there's more to it. The high school students I taught remotely also complained about time warp—they barely knew if it was day or night. They found themselves awake and on social media at 2 a.m., took naps between classes, blackened their Zoom screens to hide the pj's they were still in at 4 p.m.

For one, every day was pretty similar: put away dishes from the previous night, clean counters, check to see if Walmart has toilet paper, get out the crayons, or do Play-Doh, or go outside with the kids, prepare for classes, clean counters again, hold Zoom class, eat something, check the milk and yogurt supply, hold another Zoom class, feed and bathe kids, read to kids, sing a lullaby, wash lettuce, make rice. I was involved in each of these actions and nowhere else for those moments or hours. I was surrounded by the present; the present moved around me.

Given the location of our eyes in the front of our head and their limited number, just two, we're wired to look ahead and move into the future. Horseshoe crabs have ten eyes distributed over their bodies; scallops may have 200. A duck can see 360 degrees, a cow about 300. For them the world is ever present around them and they move within it, sometimes languidly as cows do, sometimes winging within the sky's embrace. Instead of the habitual *I have to get there impulse,* (which is physiologically male), I had an *I'm already in it* sensation (characteristically female), time during the pandemic having a gauzy sameness neither blinding nor enlightening but ever there. Vision in any moment—the fingers of my granddaughter curled around the cap of an acorn—was acute; the cumulative moments, an absorbent, non-descript hue.

The tempo changed too, in contradictory ways. Compared to other creatures in the animal kingdom, humans have a moderate ability to process visual information, something called critical flicker fusion frequency, or the point at which a flickering light appears steady. Our CFF is 60, that is, 60 frames per second. This reminds me of making movies as a kid with a pencil and a pad of paper, drawing a stick figure (usually a horse) on the top corner of each page, each drawing adjusted slightly so that the horse completed a stride, galloped and cleared a fence. The production was laborious, but then came the fun of flicking through the pages and watching the figure in action, in double time or slow-mo. That's how I imagine our CFF. As Ed Yong explains in his book *An Immense World,* different creatures have different CFFs depending on how they live in the world: a pied flycatcher's is 146 because it needs to catch insects in motion; dragonflies and flies outdo that, with speeds ranging from 200-350. A massive leatherback turtle roaming the ocean for thousands of miles absorbs a glacial 15 images per second, looking only for its staple of jellyfish (and tragically mistaking fishing nets for food). These wildly different speeds render our own sense of time arbitrary and absurd, since others' sensation of the passing of time is equally valid and occurring at any given moment and may plausibly be influenced by the rate at which a creature visually processes the world. To a sea turtle compiling images like my homemade movie, released in time lapse, a deep-sea diver swimming past would look as fast as a cheetah; to a fly receiving hundreds of images a second of me flailing at it with a swatter, l would seem to lumber like a grizzly, easily avoided, while it focused on the more crucial task of devouring my toast. For me the pace of a given day slowed when I watched a hummingbird buzz close to my face before dipping its beak into a dahlia, or listened to a chipmunk rustle under the porch and whiz past my toes, or felt clover flowers underfoot scattered over the grass. Yet the cumulative collection of days was fungible and fleeting.

I didn't go anywhere or do anything unexpected. I was simply in motion in the lap of each day: someone needed to eat, laundry needed to be done, trash had to be taken out. Still, certain scenes punctuated that continuum and became memorable. My granddaughter asked *why are there voles?* as she peeked down a hole beside withering stalks of phlox. We played basketball with a beachball and a laundry basket; danced to the Beatles and Haim; became dragons, hamsters, mice, avocadoes, traveled via magic horse to lakes in mountain craters past the stars.

There seemed a natural coupling of what we might want to remember and what we'd rather forget, a flattening of memory that denotes linear time and a sharpening of perception that sanctifies the moment. A phenomenon which, if painted, might be the scene outside my window where morning sunlight wraps around the striated bark of an oak. From beyond the horizon fitful gusts fill the highest limbs with breath so they heave and sigh. A cardinal alights abruptly, perches on the forking twigs of a white pine, and reduces the trees to an indiscriminate, two-dimensional canvas for this spark of red, this singular moment. Red beak and bandit's black mask, red wings tinged with ash.

When my dog raises his head for a look, I remember that to him, the bird is dark yellow. He doesn't see red and the cardinal is not inherently red. It appears red to me due to the construct of my eyes, the number of cones in my retinas, and the neurons that weigh the signals they receive. I try to find this enriching rather than sad.

Since I was a child I've been told that the silver grains of stars that I see are not what I see, that the light was emitted billions of years ago, that the star may have imploded or exploded. I didn't quite believe in the idea of ancient, orphaned light, but can more readily accept now the illusory nature of what I see and the species-driven relativity of how I see it. Still, I can't fathom large clumps of time—such billions, nor forty-two years of marriage, even. A touchstone occurs when my husband mentions one of my long-lost high school friends, and I start, suddenly realizing, you *knew* her, wow. But between such handholds lie ravines of indeterminate depth, and rivers snaking from unseen sources to diffuse in the sea. Salmon can recall the smell of natal water upstream and find their way back over thousands of miles of ocean, but how do they measure the two or four or eight years in between? Or is there no need—and so they don't—in any schema recognizable to us?

During the pandemic there was always the paradox of a day completed and nothing completed. Memory created and memory dissolved. Like sugar and pink and blue dye effusing from Lucky Charms into the milk where it floats and gets soggy. It didn't matter or I tried not to let it matter, giving way instead to the always-ness that arose from the ordinary and the validation of life ongoing. One rainy day I cleaned out an attic closet and found a box with an intricate floral necklace of baby sea pearls once belonging to my grandmother. I had intended to have it repaired for my daughter but had forgotten its existence. There, too, in the murky light my dark blazers and dress pants for work hung like the folded wings of owls. Or like the trappings of someone I once knew intimately.

In 2023 I wonder if using pre-pandemic things will set the mechanics of my watch chugging again, if they will cloak my existence in forms and colors that feel right—alive, even. I don't think so. I can't attribute to them that much power when for a few years I didn't miss them.

Sitting on the subway under Manhattan streets, I stare across the aisle at boots cracked by ice and wear, unlaced neon Asics, black pumps, Reebok retros, ripped flip-flops, net shopping bags, and I listen to fiction through my earbuds. I need to be there in a half hour. At 59th (where I have to switch) I skip down a bank of cement stairs, hear an ominous rumble, my feet drop faster, descending screw-studded treads blurring into a sheer incline—slip into the closing doors of an express, glimpsing the overhead sign that says the next #5 is fifteen minutes away. Score. So fulfilling. A win. Gone in the up-whoosh of the train. Sheepishly, I scan the eyelids of passengers on their phones who, naturally, don't give a shit.

At 42nd Street, passengers snake along the station platform toward the stairs. I think of a school of minnows moving in unison, sensing one another through pores along their flanks called the lateral line, (which is why they don't bump into each other), but we are nothing like them, having no common destination or purpose, no predator in common to evade, confined only by walls until we disperse under the turquoise sky crowning Grand Central. Like a gawking tourist I tip back my head at Pegasus frozen mid-stride, and there, Pisces, two numb-eyed fish loosely joined by a loopy line. And just ahead, dead center, the clock, a bald brass head with four bulging opal eyes and four sets of lordly hands. A flat voice announces delays on the Hudson line and crowds gather under the billboard, arms crossed, stalled.

It's time:

For bar-tailed godwits to set off from Alaska and migrate to New Zealand

For my second granddaughter to get teeth

For raking

For someone my age to consider she doesn't have much time left

For our ethics to catch up with AI

For saving endangered species before resurrecting extinct ones simply because we can (maybe)

For the tide to ebb

For a rat to tunnel through a garbage bag on the sidewalk outside my apartment

For the exoskeletons of crickets to crack open for the first of eight times

For a new cocktail

While the flock of would-be passengers waits, eyes on the billboard overhead for the redemptive flicker of a track number, behind them high in the station's arched ceiling two birds circle and swoop. They veer up to a ledge just under the ceiling, then flap fast and glide to the window over the doors onto Vanderbilt Avenue. I hurry up the stairs (off course, out of my way) trying to identify them, as if that would reveal a clue to their startling, thoroughly unnoticed presence. Just as I'm under them, they alight, showing me their white underwings tipped with brown. I can't tell if their speed signals ease or panic. I can't distinguish their heads or the precise shape of their beaks since anything for me at that distance looks soft, as if covered in velvet or moss. How long have they been here? How will they get out? Will they *ever* get out? They sail past filmy windows and dodge the bulbous clumps of chandelier lights that burn day and night. Night and day.

At 3 a.m. after a cop removes a man sleeping on cardboard by gate 38, and heel clicks become so seldom they ring grotesquely across the expanse of the main hall, do the birds come down for bits of fallen bagels? Do they hop about and pick unidentifiable things from concrete corners? I wonder what migratory cues batter them as they eat and shit perforce in our locus of travel.

There has been a fire on the tracks somewhere in Westchester, and a disembodied voice announces indeterminate delay. Nearly everyone shifting foot to foot below the billboard is on their phone. They will miss dinner. Netflix. A few hours of sleep. And the day will become a sort of marker: that was the day of the long delay.

———

In March 2020 the Long Island Expressway, normally sizzling with fumes and tightly beaded with cars, was wide open like a highway across eastern Montana. Just lanes and lines, like pages of a story to be written, punctuated by delivery trucks now and then bringing food to some people somewhere. Cruise ships hibernated in harbors. Fewer jets spread CO_2, NOx, CO, Sox, PAH, and PM across the heavens; fewer jets left contrails to section the sky.

Pandemic time was like living in a caul without a due date.

Time:

For a second cup of coffee

For elephants to recognize other elephants and know where they've been and when

For making curry out of leftover peppers and a half onion and frozen peas

For watching *The Wire*, again

For my husband to convert a desk into a bench

For female lobsters to pee on the males, which males find hot

For tossing out single socks

For pancakes

For a rattlesnake to bite a rat and track it till it dies and eat it

The world held its breath and waited. Waited for delivery.

As women do, measuring out 40 weeks in nausea and heartbeats and kicks and excitement and checkups and fear. Birth and death are obvious ways to clock our existence. But any birth is intimately involved with gestation, which for a mouse is 19 days, for an elephant 650, for us a non-linear span of sleep loss, cracker consumption, skin stretching, sanctity, and discomfort. As for demarcations of death, I can say my mother died in 2013. I can say I was teaching a class, though I don't know what day of the week it was, when I saw the flash of my phone and knew, yet continued the class, just as we always go on with the banal (pay bills, buy food, clean kitchen) in eerie and unspeakable ways when a lover leaves or family member dies. Or when tallies of the dead greet us as black numerals on a screen each day year after year, and I remember the numbers because I'm good at remembering numbers, though I never knew the dead and so have no way to remember.

=====

Years out, I sip coffee with oat milk from a paper cup and scroll down like so many others hunched over their screens as the subway whines and jolts.

It's time:

For Biden to meet Xi

For my granddaughter to Velcro her Nikes and head to school for six hours

For egrets to leave the wetlands

For coffee with Alina who's coming north for two days

For the dentist, dermatologist
For tree planting in Nepal
For squirrels, acorns in mouth, to dash at oncoming cars

Lying in bed this "post-pandemic" morning, I preview the day, seeing tasks and events like buoys bobbing on green-gray water that ends evenly at a horizon weighted with slate and violet clouds. I think about the novel I listened to on the subway, a story about a man whose son becomes a coke freak and abandons his wife and kids, finds god and comes back, does social work instead of selling cars, and talks to dysfunctional kids but can never get through to his dad. I resurrect scenes of the people I saw down there underground: a burly guy hefting a saxophone; a woman with an orange-emblazoned Target bag who stood up to let a child sit; a man whose jeans dragged on the linoleum as he passed by, cup in hand, recounting his story in one car-length sentence after another, moving so fast the guy next to me who reached for his pocket couldn't pull out change quickly enough before the man disappeared.

AIR

Wind

M Y MOTHER USED TO say she hated wind. I've found myself
saying it too, whether from internal iterations of her or from gen-
uine dislike, I'm not sure.

But today there's a soft pink wind, a wind that films around my body
just as sweat balances at the pores on a humid August morning before
the sun has heaved itself high in the sky and melted the clouds. The wind
is soft cotton, coral pink, soundless.

I learned two days ago that my daughter is five weeks pregnant. Two
cells, locked together in an embrace more intimate and enduring than
any conscious one, adrift inside her, somewhere, gently moving like this
pink wind. I can't see either. Both remind me of life continuing with or
without me.

I'm watching sassafras leaves jiggle and spears of salvia waver as bumblebees weave among them stopping here and there for fuel. The upper
limbs of oaks shift indecisively. They don't reveal if the wind comes from

the north, south, east, or west. They don't conform in any way. They signal a presence only, a slight push of molecules of air.

We wouldn't know if spirits moved among us, or a god massaged the petals of a peony. We wouldn't recognize the divine, how could we? I can't see the wind or the beating of my heart or the making of a heart in my child's uterus.

I wonder about the dependency of embryo to uterus to mother, of wind to sun and the earth's rotation, of the divine to our conception.

If the earth didn't rotate, the wind would blow north and south from the poles. If the sun didn't heat the air, there would be no pressure differentials, and cool air would sit static and languid as an anchored ship. If we hadn't poured carbon monoxide into the atmosphere, breezes could play on the streets of New York, and I wouldn't hear warnings to keep your child inside, don't run. Air quality hazardous, clogged as the Uber-clogged streets.

If my daughter had not so passionately wanted a child, this embryo would not be. My husband worries that if she runs on concrete it will fall out. I disabuse him of this but wonder about other reasons it could.

At midday the wind blows a bit brighter, almost green. I don't know where it came from or how far it flew. All I can count is time, time till I know the pregnancy is not ectopic, time to find shade, time to watch the clematis that outgrew its frame so that now two feet of vine bearing shiny leaves and white flowers veers off at a horizontal angle in search of support, the thin tip dipping despondently. Where to go from here?

I go to the hardware store and buy bamboo stakes and green twists, and concoct a makeshift extension to the frame on which the plant is growing. I don't know if the plant is any happier, reaching vertically to the sky, but it doesn't look so vulnerable and unsure.

Today the wind is translucent blue, open and undirected as a newborn's eyes, my daughter's eyes, before they turned dark as walnuts. Hot air over driveways and back decks has lifted, leaving a gateway for cooler air from Peconic Bay to glide in, poised and unannounced, amorphous yet steady, steady as the flap and shush of small waves on pebbles and empty clamshells. I don't know if any language has words that precisely mimic this sound.

In the city my daughter is gazing into a screen, looking for her child for the first time. A pomegranate seed, a lentil, an eraser tip. Did it travel to the uterus as charted or get stranded in the channel en route, abruptly becalmed, snagged on a scar? Does she see dark shadows where its eyes will grow, the spot where a heart is beating in quick silent iambs twice as fast as hers or mine—to live, to live, to live?

Once I dreamt that I floated somewhere outside the earth and looked back at an azure globe, inner lit, luminous, loosely swaddled in clouds. More than anything I remember color (have I ever again dreamt in color?) stately blue like the wind today. Blue that gives nothing away, reveals nothing, is not uncalm. Little embryo. Who was I to glimpse it? And had I had words, they would have vanished the instant they formed, swallowed by space where such things have no life.

───

I used to curse the wind when it spun off the Hudson and knifed between high-rises as I trudged west to my apartment after work. It would be early evening though already dark for three hours as I wrapped a scarf around my head and across my face, tightened my fists in my pockets, and leaned against the muscular hand raised against me. The wind felt like the sound of my boot heels scraping and crunching on salt. It carried all that energy on so many nights whose days under electric lights I don't

remember. It blows like that still though I'm not there to record it. It
blows though I don't bother to find out why.

My Ithacas have shifted. The steady west wind (that had pummeled
me) brought Odysseus so close he could see cooking fires on his island, so
close he shut his eyes and breathed the yellow wind and drifted, dreaming
of his son and wife while his men opened the precious bag that Aeolus had
given them. Treasure! they imagined, dipping in their hands and feeling
not the cool touch of gold and jewels, but air, dark violet fistfuls of air,
furious squalls that blew them helter-skelter like so many fallen leaves in
November when the wind swirls them into a column rising, splintering,
peeling outward like the outer wheel of a cyclone. So we wander, inadvertently. So we find other Ithacas. We've long forgotten the first, that slow
vital landing on the uterine wall.

———

I don't know why my mother hated wind and now she has died, I can't
ask her. Maybe it messed up her hair, which she had set every week by
the same hairdresser in the same style. Maybe she found it invasive, deaf
to her wishes.

Erratic winds this summer are feeding wildfires in Mendocino that
leap across creeks, roads and fire lines, rage three hundred feet in the air,
refuse to settle at night, and voraciously gobble tens of thousands of acres
a day of dry timber and brittle grass and homes. Winds reach 140 miles
per hour, fueling fire whirls, which suck air at the base and rocket into
vertical columns that strip bark off trees, pluck them out of the ground,
toss firetrucks like toys. Winds pick up embers and spit them miles ahead
of the fire. Experts say the fires will burn all summer, and ash-laden wind
is coating the state with carbon dioxide, carbon monoxide, chemicals, and
soot. Shots from space show fulminating gray clouds chaotically growing

and enclosing the earth below, an earth already charred dark chocolate and missing visible evidence of life.

Here on the other coast a taciturn sky and monochromatic cloud cover obscure the horizon. Faint rain hisses in the leaves. The static morning offers no information about the rapid formation of liver, lungs, feet, or the rapacious destruction of sun-hot leaves, no hint of process which is life force, life force which is process. Silver wind. Cool and indifferent. Wait, it says, wait.

⸻

My daughter's body is not subject to her will (or mine) to safeguard the fetus, to not feel acid skinning her stomach, to not bleed, to determine gender, to carry the baby high or low, to deliver swiftly, safely. The body is its own engine now, and the fetus both dependent and terrifyingly not.

There is never that silence of the womb we imagine when subways screech to a halt and off we go to another day. There never was. A fetus of just over four months hears mass transit in action: blood pumping through its lifeline, a huge heart beating, stomach gurgling and twisting as enzymes go to work. My daughter's baby will hear air soughing in and out of her lungs as she sleeps, waterfalls of air as she races to catch a train, green zephyrs, orange gusts. Molecules of that air will travel to the baby through a single vein.

Tonight thunder explodes directly overhead, knocks out the power, sounds like a bomb. The dog breaks through the screen door and barks inanely at the night. I'm touched that he thinks he can protect us. I dash from window to window with towels as rain puddles on floorboards and bleeds across rugs. In the morning I learn that two inches came down, but pine needles hang listless, slimy petals strew the deck, air refuses to move— everything is in conspiracy to conceal the violence of the night—while high in the atmosphere mother-of-pearl clouds glide against a steel backdrop of clouds whose depth I can't measure. They lighten the sky here and there but leave no wake in the darker gray that stays. Their movement is scarcely perceptible, the only perceivable sign of wind. It is not the silence of life beginning, though I'd like to think so, but the gracious silence of inevitability as they scud one place to another.

Life Inside

THIS MORNING MY DAUGHTER asked me what it feels like when a baby kicks in utero. She's pregnant, and her progress reports forecast some tangible action in the coming week. Is it like a cramp? she asked. No, nothing like that. It will be far more gentle, discrete, and clandestine than the circumstantial evidence of new life to date: vomiting, mind-sucking fatigue, vulnerability, and fear. No, I thought later, that small push feels the way the nub of a crocus looks, pushing through just thawed earth patched with wet oak leaves from the previous fall. A round feeling, like the mound of a tunneling mole, vulnerable and undeniable. You will know, I told her.

I can't help but follow the life of my future granddaughter in a way I never did with my own children, when I simply watched an expanding belly in the mirror and tired of the rotation of three wearable items of clothing. Ease of viewing in-vitro images online aside, I find myself astounded at the creation of kidneys, bones in her ears, rapid-fire heartbeat, emergence of

taste buds and eyebrows and eyelashes in a fetus that weighs three ounces
at this point and my daughter not visibly pregnant. It's all underground.

———

I remember the nudge of a tiny fist or foot months before the insistent kicks when the baby was getting ready to exit. Why did she choose
this moment? I would wonder. Was she asleep, dreaming, waking? Was
movement instinctual or reflexive, did the world jar her suddenly? As my
daughter reads about a synagogue shooting one day, night club slaughter
the next, about starvation in Yemen and incendiary vitriol from every
political mouth, as she drags herself to work, fretting about what her
boss will say when she learns about the pregnancy, I tell her she is only
sending anxiety surges to her child. She should sing to her. She has that
power, also hidden, dormant as a milkweed seed under meadow grass
turned stiff with frost. She smiles, yes, okay Mom.

Milkweed flowers are as complex as orchids, but I never connected
sensitive petals to the barnacled pods that burst open in November, nor
realized those parched gray pods were preceded by pinkish blooms. Today
on a hillside overlooking the Hudson, I found a cluster of naked gray stalks
with elf ear pods split open in eighteen-mile-an-hour winds coming off the
river. White fluff still clung to the rims, though flecks broke off here and
there, spun in an updraft and settled on a hillock of toppled meadow grass,
leaving a precious cargo of two or three seeds to sift through the spears
and settle in the dirt. I picked up a cottony wisp, soft as a newborn's hair.
Then, with only momentary guilt, I broke off two brittle spears, brought
them home, and set them in a jar on my kitchen table. They looked sort
of self-conscious there, as if setting the tone in a high-end furniture shop
in Soho where lights dangle from a cathedral ceiling and batik pillows
complement the sofas. Four seeds have come along, denied their parachute

ride over the meadow, but before the first frost, I'll plant them. That's the sudden plan, though I have low expectations of any germination from seeds. How could they possibly sprout all alone and so small? I usually buy trays of impatiens and vinca and salvia that someone else with expertise I don't have grew to a safe, visible height of a few inches. Nurturing has its limits.

Milkweed is not a weed but native and non-invasive. Several species are endangered. I'd like to see globular bursts of pink like so many blossoming fireworks in the backyard. But maybe my daughter's child will pluck a leaf, which contains a toxic white sap that could irritate her eyes or raise a rash on her pure, new skin. Or maybe I'll nurture a cotillion of monarch butterflies who lay eggs only on milkweed leaves because that's all the young caterpillars will eat. They tolerate low amounts of the glycosidic poisons, but their bodies absorb the bitter taste and retain it through the pupa stage and into the emerging butterfly so that many potential predators look elsewhere, warned by the bright orange wings. Milkweed thus offers food and protection, dual parental roles in one, though safety isn't foolproof. Most of the toxins lie in the skin of the butterfly's abdomen, so black-backed orioles unzip the cuticle with their sharp beaks, discard it, and devour the juicy interior. Black-headed grosbeaks have a higher tolerance for poisons and leave nothing behind. Just as parental influences wane as a child grows, the protective potency of toxins may wear off after a few weeks, leaving the butterfly more palatable, more vulnerable. Where monarchs winter in Mexico, huge flocks of grosbeaks often descend in the morning and again in the afternoon, leaving the ground littered with wings like so many rusty November leaves.

As I look at my daughter's glowing face and divine the hillock under her shirt, I wonder what will threaten her child a decade from now or three when I may be here to witness it or not. I wonder about the morality of artificial intelligence, about hopping in a car with a ghost at the wheel, about the omnipresence of eyes on every step she will take, about her questions when truth has become not relative but unstable and anachronistic, about the task that lies ahead for my daughter, still my little one, who calls me to untangle a crisis of betrayal or to tell me about her latest sonogram, or to ask if she should thaw a slice of bread before popping it in the toaster or if the yogurt that she left out last night is safe to eat.

In the later months of pregnancy, I remember glancing at my enormous belly and seeing the taut skin suddenly jolt. Not long to wait. Right there was an insistent elbow or foot. Even so, the birth of my first child, a boy, was slow and hazardous. The umbilical cord became wrapped around his neck and oxygen levels dropped; he ingested meconium; his first bed was not me but an incubator, his first sustenance antibiotics dripping through a tube. Why didn't I panic during that delivery? Why did I blindly keep going, assured that I just had to deliver him, that all would be fine for how could it not be? This was life, it was time. But of course, this is not always so, might not have been so, is not always so even for the lovely monarchs with their toxin-empowered skin.

I think about my daughter creating tiny fingernails and functioning lungs and nostrils, about the click of months and growth to come, and I think of a black-and-yellow-striped caterpillar curling up in a chrysalis to grow. In 240 hours it will split open its silky shroud and unfurl elegant wings, russet-streaked and framed in black, of which there had been no hint before. Perhaps because the metamorphosis is so extraordinary, it has become a cliché, a threadbare symbol of transformation, of ethereal beauty arising from a lowly thing. Such cause for hope.

Yet, when the butterfly struggles out of its pupa, for three long hours its wings are wet and as utterly useless as a newborn's hands. During that time a yellowjacket or paper wasp can attack and devour the butterfly, toxins and all, before the wings ever open fully, before the butterfly feels the jittery tilt of first flight.

Perfect Kill

DRAGONFLIES ARE LIKE WAZE. They predict. They don't chase after mosquitoes, they intercept them, which means they have to calculate the mosquito's distance, speed, and trajectory. Dragonflies perform these calculations in milliseconds, far faster than the Waze app, which always takes agonizing seconds, even minutes, to load as it computes the route, average speed of drivers, and quagmires that might await.

Dragonflies' calculations are more precise. They dart after flies and mosquitoes with a successful kill rate of 95% and rarely stop. They grab, decapitate, de-wing, and devour in flight. They may copulate aloft, male grabbing female mid-air, then zoom without pause after the next hapless mosquito. Their four wings can flutter independently in any direction, a complex choreography at thirty beats a second, which lets the dragonfly rise, fall, or skim side to side like a helicopter while expending far less energy than a fly, whose wings beat at 1,000 times a second. Still, dragonflies need to eat all the time, sometimes consuming hundreds of mosquitoes in the course of a day, zigzagging along shorelines and gardens at speeds up to thirty miles an hour.

Articulating an ice hockey version of dragonfly tactics, Wayne Gretzky advised, "Skate to where the puck is going, not where it has been." This comment has appeared in millions of PowerPoints in businesses across America as a call for innovation. The dragonfly's success rate sets a high standard. Maybe an Amazon warehouse with its swift conveyor belts, and precision-wired robots, and humans standing ten hours packing single-item boxes wins out. The company is seeking perfect efficiency and it's making a lot of people nervous, but soon those jobs will be gone, replaced by robots and assorted AI. We rarely see where the puck is going.

———

I don't know if perfection exists in nature, or how we would recognize it if it did. A buttery daffodil with a flaring trumpet at the center of a spiral of petals looks perfect to me, but so might another that is smaller, white, dotted with orange at the heart, so which is it, or both? We don't have an absolute paradigm against which to measure a daffodil—or dog, man, or woman—though for centuries we have conceived models of aesthetic perfection, human and divine: Poseidon poised to hurl his trident, left hand deftly raised in the direction of flight, right arm back and eternally cocked. Or Nike of Samothrace leaning into the wind with her marble dress rippling like water and wings flung back in honor of some sea battle in which humans have long been forgotten. Or David, the giant-killer, with his alert eyes and crown of curls.

———

When my son was two, I still looked at him with the glazed eyes of a new mother enthralled with the beauty of her first child, his honey hair and deep brown eyes. One afternoon we headed to Riverside Park in Manhattan, he on his little motorbike propelled by pumping his sturdy legs and pushing off on the sidewalk with his Keds. Once off the streets, he sailed ahead and within seconds slammed into a girl on a mini two-wheeler, the impact sending him headfirst over his bike and jaw-first onto cement. I raced over and picked him up, blood streaming from his mouth. I didn't dare look but turned abruptly with him in my arms, chin on my shoulder, and rushed back to the street. Once in a cab I peeked at his blood-spattered face, at his two front teeth—broken—one straight across, the other at an angle. About half of each tooth remained. When the dentist ascertained that they wouldn't fall out, he said I could have the teeth repaired but allowed that they were, of course, baby teeth. Of course? We would have to wait four or five *years,* and all that time my child would be marred. I had created a perfect being and the world was starting to interfere, chip away at snowy teeth, carve unmapped skin and unscarred knees. In subsequent years he would get gravel in his shins, break his wrists twice, suffer pernicious ingrown toenails, develop an overbite, get pimples. But that afternoon I went home and discussed the idea with my husband, who said I was out of my mind. I sighed and conceded. The teeth became one of those flaws that lends character, one could say.

The Greeks knew not to conflate aesthetic and moral perfection; their gods often behaved in scurrilous ways, subject to fits of emotion. But in the Old Testament, Moses made a staggering claim about God and his creation: "He is the Rock, his work is perfect." How can anyone make such an assertion unless one's definition of perfection includes imperfection, ranging from a child's inconsequential tooth to hunger, greed, and slaughter? But the biblical passage gives an odd justification: "for all his ways are judgment: a God of truth and without iniquity, just and right is he." All His actions are law, and that law based on opinion, which is certainly relative, but relative to a divine being who is believed to be perfect. In Deuteronomy Moses leaves no room for error or skepticism, trusting the opinion, or law, because God is always fair and correct, which is ad hominem reasoning if we can apply such a term to God. Alternatively, *judgment* can mean a calamity or punishment sent by God. Rather than being perfect, we are the objects of divine punishment, which he is just and correct to inflict, and for what? Perhaps our moral imperfection.

There is a possible contradiction here of two truths, a disconnect between intention and outcome. "Between the conception / And the creation / ...Falls the Shadow," as T.S. Eliot wrote. Like one of those drowning humans watching Noah's ark float by, I find myself clinging to a splinter of wood, still believing in the purity and possibility of creation in its most mundane and quotidian iterations. As such, the most perfect element in my life today is my daughter's pregnancy. I'm weathered enough to know she will not have a perfect child, that fractures, anger, rebellion, and tenderness mark the highway ahead like sundry billboards and neon motels, sirens whirling at accidents, slivers of the moon through an opening in the hills. It is the concept of this pregnancy, the idea that life is being created, another generation in the making, that alone seems emblematic of a paradigm that exists among any surviving species on the globe, and in it lies perfection, the natural order to which I'm witness and a part. What feeds the abstraction, oddly, is my wonder at what is physically occurring at any given moment in my daughter's body: the formation of 100 new brain cells each minute while the heart flutters 150 times, the sprouting of tiny fingers that will soon hold hers, and then another's, the sounds of my daughter's heartbeat of which this fetus knows nothing—but is listening. These tangibles construct and confirm the idea of more life, more energy, a fundamental form of perfection, even as my daughter vomits, bites her lip to keep down a saltine, drags herself from bed wondering what alien being has invaded, how she could have two hearts, if her life as she knows it is gone forever.

———

Approximately 300 million years ago, when the earth's atmosphere had more oxygen than it does now, giant dragonflies buzzed around with wingspans of two-and-a-half feet, taking out other dragonflies, mayflies, sixteen-inch reptiles called Petrolacosaurus, and amphibians. Today's descendants have the same extraordinary eyesight, which along with their flight agility, makes them near-perfect hunters. Dragonflies can see 360 degrees. Eyes encircle their entire head like a helmet, with 30,000 facets feeding eight pairs of neurons that deliver all these images to the brain. By way of comparison, a fly, also a vigilant creature, has 6,000 facets. Dragonflies detect light in the same spectrum we do, plus UV and light polarization, which helps cut glare over water. Perhaps more incredible, a dragonfly can frame points of interest and literally block out the rest, like looking through the scope of a semi-automatic to zero in on a target and hone the precision of attack.

One late August evening I walked to a beach on Peconic Bay to see the sunset as I might on any day. I was startled to see a perfect globe, as if drawn with a fine pen, staring directly at me with ochre orange calm, while over the water it shed millions of flickering glints signaling faster than I could measure, sparking and darkening as the water trembled. At the edge of the beach, backlit by the sun, were scores of dragonflies swirling and diving and rising in what struck me as a mating dance—but was, in fact, a kill. A swarm of insects had just hatched and the dragonflies were dancing, eating, devouring in what I felt to be the counterpart of my daughter's creation of life. Here was perfect death. Sunset and all.

The next afternoon I was reading by a pool when I looked down beside my chair and mere inches from my foot, making its way through unruly spears of grass, was a huge black spider with a thick abdomen and hairy, arching legs, the largest I'd ever seen. Weirder still, once it made its way to the bluestone, I saw that six inches behind and seemingly attached to the spider's single filament was a half-inch black insect with wings and a curved abdomen like a hornet. The two, spider ahead, insect following, made straight for the pool. A few inches from water, the black-winged hornet pounced and the spider went dead still. I waited seconds, minutes. The predator stabbed at the spider, eating it, I thought, and my sympathies turned. Poor spider. I crept closer to look and suddenly the spider flipped over and made a dash for the pool, the hornet buzzing after it in hot pursuit, both tumbling over the edge and landing in the water where the battle continued. By now I felt too sorry for the spider and ran to get a skimmer to lift out the battling pair and put them on even ground. When I returned they were nowhere in sight. I saw no bodies, the water was clear. Suddenly over the lip of the pool the hornet appeared, dragging the spider, which did not move. At this point I could predict nothing, it was all new. The spider had turned ashy gray and lay on its back while the hornet covered its curled body with its own. As I looked more closely, the hornet spooked and took off, abandoning the spider on the bluestone. Should I flick it into the grass? Why did the hornet chase it down, stun it, heave it from the water just to leave it there? Some minutes later I looked and the hornet had returned, victorious, and dragged away the spoils of war. There was no trace on the quiet stone, as if nothing at all had happened. I was stunned. What had I witnessed? A vicious, prolonged, dramatic battle that shattered my notion of death as the closing parenthesis that balances the opening one, death as in any way perfect, for a fly, a spider, or me. An unpretty sight, morally indifferent. Reason says it must be; my heart balks.

Somewhere under the trees, perhaps nearby as I admire the sunlight and flickering monarchs over the grasslands, this tarantula hornet will lay an egg in the safe body of that spider, paralyzed and still not dead. The young will hatch to a bonanza of food. It will devour non-vital organs first to keep the spider, its lifeline, alive until it flies off on its own, energized and empowered to kill and live.

Criminals We Know

THE BRITISH CALL SEAGULLS the "thugs" of the bird world. They "detest" the birds for snatching food from picnic tables and depositing splotches on cars. Pigeons and magpies come next on the list of most loathed, along with the unlikely sparrow who is simply "dull-looking," a criticism that seems specious coming from a populace of oxfords, woolens, and tweeds. That the seagull should be so abhorred stuck with me when I read the posting on UK's *Daily Mail*. I cannot imagine *hating* a seagull.

But I have feared them. Brian and I, once while blithely exploring a rocky shoreline of Block Island in our courtship days, happened upon a massive colony of gulls. They set up a raucous alert as we tiptoed over stones, scaled boulders, and stole around the ashy birds that stood erect and started toward us with the singlemindedness of a mother elephant. Surely there were eggs and young. We watched where we placed our feet as heads cocked in our direction and adults brandished their wings. *Keow, keow, hahaha* reached a screeching crescendo. A single bird struck me as inquisitive, a colony, alarming. Was it the numbers, like rats in *1984*, or crows in *The Birds*? They could dive-bomb my head or nip my bare toes. Seagulls sometimes land on surfacing whales and take a bite out of their backs, but I didn't know that then.

I'd always associated seagulls with open horizons and freedom. At age eighteen, with high school and parents three thousand miles behind me, my two best friends and I sprawled on the deck of a ferry boat winding up the inland waterway from Seattle to Juneau. In our army pants and bandanas, with our single-lens cameras and journals, we watched seagulls cruise in our wake and veer down after a hot dog roll or bit of paper. I thought the gulls skimmed on air currents behind the ferry, linking us in some ethereal way, and I mused that, like us, they didn't care how far they traveled or how far from anything they were, especially as their white bellies and black wings glistened in the sun long past the time the sun should have set.

As a kid on the Jersey Shore, I'd race up and down the beach with my brother and ride waves on a rubber raft. Gulls were simply part of a backdrop, strutting with their stiff necks, winging for a few yards when I chased them. At any beach they might be bobbing on the water or hanging alight on a solid wind, white bellies reflecting whitecaps flicking across the water like accents on a gray blue canvas, an element of design.

One frigid day in January, I walked along the shoreline of Peconic Bay. Even the sky felt apathetic, the beach littered with pecked whelks and empty slipper shells. No *keah keah keah* punctuated the air. No boat churned past. Small islands of ice scarcely moved. A lone seagull suddenly whuffed overhead and dropped a clam just twenty feet or so before me. Down he swooped and grabbed it, flew up, dropped it again, *konk,* dropped down, picked it up and rose again. I followed him, happy for the company, and impressed by his reasoning and use of gravity. He was not a decorative bit of icing on the seashore scene but a bird with intent—one word left on a blackboard otherwise erased. He was cognizant of me and undeterred; when the shell cracked, I left him to it.

Way down the beach I could just make out a gray-flecked gull pecking at something in the sand. As I approached, it kept at it, not flicking an eye in my direction. At its feet lay the dead swan I'd seen the previous day. I had stopped to stare at its huge crooked wings with long, disheveled feathers, its black legs and gnarled paralytic feet, its ribcage stripped bare, its head gone. I had tried to keep my dog away. Now a gull picked at the corpse though nothing remained inside the ribcage, and the bones lay there like the frame of a hand-hewn boat.

Where were the mates, the colonies, that winter day? Were these two left behind, like a Canada goose I found there another time, wandering alone, calling and calling. Quick research revealed my ignorance: yes, most seagulls migrate. More startling: seagulls per se do not exist. If I had looked closely over the years, I might have noted a red beak or yellow, pink feet or yellow, a black head or white. Some sixty species of gulls are members of the Laridae family, which also includes skimmers and noddies and fairy-terns. Some gulls don't even *live* by the sea. But since gulls tend to be some combination of white and gray and black, we clump them together. It's like saying that everyone east of Israel is Asian.

Great black-backed gulls have wing spans of five feet. Franklin gulls migrate from North America to Chile and Peru. Ring-billed gulls have just that; black-headed gulls actually have dark brown heads that turn white in winter. Gulls themselves know something about individuation. They mate for life. And for some, the courtship is intimate: when she's ready she peeps *klee-ew* like a chick and an interested male advances, uttering *huoh, huoh, huoh* (I'm here to stay) and regurgitates at her feet. If his offering appeals, married life begins. Habits and diets among gull species differ, too. Laughing gulls that breed on Long Island migrate to Central America and sometimes as far as Peru; Bonaparte's gulls capture insects in the air and pluck surface fish with stunning agility, foraging inland on the Great Lakes or miles offshore where the Labrador Current meets the Gulf Stream. The largest, most aggressive gulls, the black-backed, eat just about anything from human garbage to bird eggs to carrion.

Does that make this gull, or others by association, a thug? How did it become one? Maybe its nest was bulldozed, its shoreline covered in concrete. Maybe oil from a single jet-ski seeped into the shallows matting the delicate barbs on its protective feathers, or pollutants spawned algae that deprived fish of oxygen. Naturally, a gull that finds no minnows or mollusks will grab a sandwich, tear through Glad Wrap, steal from other birds, steal from each other. Who wouldn't? But this is a liberal stance. The opposing party would argue, the bird should not expect free lunch.

Given the omnipresence of humans, you're more likely to find a gull stabbing the tuna sandwich in your picnic basket than eating a dead swan. Such theft suffices to elicit the hatred of the Brits and the admiration of the Irish who, as underdogs, appreciate the gull as a trickster figure. Their Celtic god of the sea, Manannan mac Lir, sometimes pictured as a gull, owned a boat called *The Wave Sweeper,* which moved without oars or sails to shepherd souls to the afterlife, often in a cloak of invisibility. He's also identified with another trickster, Bodach an Chota Lachtna, which means "the churl in the drab coat," recalling the gray and black plumage of a gull. In grass and meadows gulls may do a rapid paddle dance that issues vibrations underground, mimicking the sound of rain. Worms can travel to new territory faster on the earth's surface than through the dirt, but since they need to stay moist in order to breathe, they only risk surfacing if the ground is wet. The gulls' quick steps, called worm charming, lure the earthworms upwards where they're quickly devoured. This trick is not innate; with each new generation the gulls need to teach it; the worms never learn.

In the dead of summer, families with coolers and umbrellas trundle onto the beach and set up camp. I walk past, watching them smear on lotion and lie prostrate in the sun. Dozens of gulls cluster on a stone jetty about a hundred yards down the beach. Waves sigh and seethe over the rocks, bringing in crustaceans and plankton, and the gulls are busy. Food is abundant, little ones thrive. A burly man and his son head toward the jetty, fishing rods in hand. On a plush blanket a couple embraces, he leaning over her, she lifting her head and turning toward him, the top of her bikini top undone—and as he strokes her back and she sees nothing behind her Jackie O shades and he's thinking about the night and she's thinking about her tan, a mid-size, round-bellied gull hops toward them, cocks its head, its wide-set eyes like bits of obsidian, struts, waits, hops and without a blink, snatches the just-opened sandwich, parted and fleshy in the sun.

Brian and the Snowy Egrets

M Y HUSBAND DIDN'T DISCOVER nature until four lithe lion-esses were striding beside our jeep in South Africa not ten feet from his right hand. It was dark, and we watched their tawny shoulders and sinewy hindquarters pace unchecked along a dirt road and into the bush, where we followed, anticipating their evening hunt, which I both did and did not want to see.

Before that formative journey Brian didn't know a leopard from a cheetah, a robin from a crow, an oak from a maple. Now he stands on our deck looking out at the wetlands of Peconic Bay and gushes, "Look, look! It's the snowy egrets!" And now he wants a bird sighting scope for his birthday. Yesterday I told him that snowy egrets have yellow feet. He put down his crossword puzzle and looked me right in the eye, asking "Why?"

"I read two accounts—one said the feet only turn yellow when the bird is ready to mate, the other didn't qualify the yellowness."

"Well, which is it? You should find out."

The snowy egrets wing like angels over the phragmites behind our house. They alight and stand very still, watching the water, a fine paint stroke of white against the wet greens of spring and dry khaki of fall. Brian believes there's a pair, but I don't know how he thinks he can identify them from a distance of a few hundred yards. As far as I can tell he believes in monogamy, so maybe he's projecting onto these creatures he clearly admires.

Like Brian, the snowy egret is not a fussy eater. Without complaint, my husband eats day-old salad whose oil has congealed the sodden leaves and robbed the snap even from Romaine; he peers at a six-day-old chicken and shrugs, "Smells okay." The delicate egret will stab a lizard, bite at it till it's dead, and swallow the creature whole; so fine is the egret's neck you can follow the bulge as the lizard descends. A snowy egret consumes snails, frogs, crayfish, worms, and mice. Standing two feet tall, with a wingspan of four feet, it weighs less than a pound.

My grown kids scoff at their father standing on the deck at night sighing over the view. "There's an absolutely *amazing* moon tonight," he exults while they fail to stir from the table and pour some more wine. Whenever he's cooking, he dashes from the kitchen at three- and four-minute intervals to catch each stage of the sunset.

"Yep, Dad, the sun did it again."

I wonder if it's a function of age. If we look at the frontispieces of Walt Whitman's *Leaves of Grass*, which appeared in nine different editions over a period of about thirty years, we find first the young braggart, head cocked, hand on his hip, staring at us head on, as if he's got something to tell *us*. So, too, with a woodcut some years later, a headshot only, with bright engaging eyes that invite *us* to listen (despite Whitman's caveat that his eyes were actually dim and lazy). Later we see the old mystic poised in profile, butterfly on his fingertips, looking off into the sky, past the butterfly and its probable flight, and far into the ether. By then he was weary of politics, disillusioned with rifts and corruption in the country, tired of talking for us and to us.

Brian, a journalist and former extravert, traveled a commuter train enough times to circle the globe two times (arctic terns migrate the equivalent of four round trips to the moon, I tell him), ascended elevators to his office on the twenty-second floor, looked out un-openable windows, and breathed recycled air. Now he sits on the deck and stares, just as his father, a retired electronics engineer, used to do from his porch on the New Hampshire coast.

"What are you thinking about?"

"That tree there," (I crane to divine which one among the morass of scrub oak) "see how it's leaning, I'm thinking of taking it down. Catch more of the winter sunset."

I'm glad Brian discovered nature, and I like watching how he likes it and what he likes. It's not the impassioned conflation of soul to tree trunk or branch angle that comprised my teenage years as I trod forest trails, hiked mountains, swam in chilly lakes, and traced the tortuous heights and depths of my existential self along a jagged horizon of peaks. Thank god, it's not that. It's more selfless, either because he was never one to introspect or because he figures whatever he is, he is. Still, those snowy egrets touch something (like the brush of a certain hand on your arm) or whisper that no matter how long or how often one witnesses an afternoon sun turning feathery seed heads white gold as it glances through salt meadow grass, purple loosestrife, and bulrush, that phenomenon will evanesce, marking one fewer in the finite schemata that is your life.

"It turns out the egrets' feet are always yellow," I tell Brian. "Greenish yellow during most of the year, but bright orange yellow when it's time to mate. And the bare spots on their faces turn red."

"Hmm."

"The courting males squawk a lot and pump their bodies up and down and even perform aerial acrobatics."

"Cool."

End of sex conversation. He's Catholic, after all, or was. When Brian was growing up, his mother didn't even tell him when she was pregnant, which was frequently. Four times he thought she was getting fat, and then she'd disappear for a day or two and return with a crying thing to feed and wash.

"Some things you kill by talking about them," he says.

So, as he sits on the deck looking past the railing and past the blueberry bushes, oaks, and white pine to the glistening wetlands and flank of land across a sliver of the bay, I keep silent.

One Christmas I gave him two bird books, abundantly illustrated and informative, but he never opened them. Brian knows an osprey from an egret, but not much beyond that, so the fascination may not be with the biology of birds but with birds in the context of something beyond themselves. Out here the light is brittle and the shadows severe. When clouds roll in, a hawk sailing over winter trees seems an inconsequential piece of monochromatic gray, but when afternoon sunlight catches its flecked belly and sets bands of forest floor ablaze, I can't *not* look at the hawk's imperious circles.

Just as, every time I walk along the sound, the water commands that I notice—one day molten steel, another day indigo, sienna, blue, gray green like an old Coke bottle turned to sea glass, silver, bruised, and black. Light transforms the water, moods, us. We're subject to the wavelengths we see and miss so much more. These frequencies, light energy, sound energy, are congruent in ways I don't question or fathom and dissonant in ways I want to understand. That quick Donald Duck garble and raucous cough come from the ethereal egret. Those high-pitched whistling chirps in rapid succession and tremulous squeals radiate from the beaks of osprey with wingspans of over six feet. I don't think Brian deconstructs as I do. Nor is he passive. He casts something of himself over a field turning russet and ochre in the afternoon sun, but I don't know what that is.

"Does Dad know there are other views?" asks our daughter as once again she catches her father sitting alone, staring at a purple cloud. "I mean, it's funny, weird."

———

One night, after the dishes were done and Brian had fallen asleep, I noticed a strange shadow across the side deck. I peered tentatively out the door. A great blue heron was on the railing not five feet away, cocking its head at the fluorescent-lit kitchen counter. Then it watched me watch it. This might have gone on for a while except my son's dog charged the door, slammed into the screen, the heron raised itself on elegant legs, opened its vast wings, and with breathy beating rose into the air, quickly vanishing in the dark. It left behind a ghost of itself that I saw again on the railing and that I still see even when not in the house, as if something other than the bird inhabited the bird, as if something other than the bird was the bird, not a spirit or god incarnate, nothing like that, but not the bird that alights in the wetlands and stands waiting for prey.

When I told Brian about the heron and its untimely visit, he said, "There was a racket the other night. Raccoons on top of the garbage shed. They got in, too."

"And what'd you do?"

"Shone a flashlight on them. That didn't deter them. Finally I yelled and they quit."

I looked at him as quizzically as the heron had looked at me, head on and unblinking, and continued to talk, anecdotally. He marveled at nimble raccoon hands and raccoon determination while I stirred my tea and remembered a bit in the news about a baby caught in a fire.

But the heron, I want to say, a great blue heron, just there. Had he seen it, what would he feel? We see colors differently; he watches TV at decibels that rattle my nerves. Had he seen it, what would he have said? Would he have awakened me to see it? After decades of marriage I should know such things. When I close my eyes, I see the misty blue heron cocking its head and its audacious eyes. Why was it here? And why are we? We never expected to own this house perched on the hemline of wetlands and feeling that the heavens perfectly arc above our roof. When Brian stands on the deck looking out, I want to ask, what does the moon say to you tonight? But to ask feels like an invasion, as if I were inquiring about a lover, which, like a lion kill, I both want and don't want to see.

The following morning, coffee mug in hand, Brian calls out, "The egrets are just there, take a look!" I stand beside him on the deck. We see four standing erect as another glides just over the marsh grass. For a while we stare, my bare feet warm on the boards of the deck.

"Did you know they were nearly hunted to extinction in the late 1800s?" Brian looks surprised. Either he didn't know or he's surprised that I do. I tell him the birds were shot for their plumes, which were a fashion statement on hats and gowns, selling for $32 an ounce, which at the time was twice the price of gold. I'd googled the egrets one evening and liked having some facts.

As Brian pulls out his binoculars to get a better look, he says, "If things continue at our current rate, water will rise five to ten feet by the end of the century, erasing New York City, Philadelphia, Boston, San Francisco, LA."

Six months later extremely low temperatures hit the East Coast, freezing the sound nearly halfway across to Shelter Island. I walk alongside thick, crusty ice marked by hillocks and crevices. It seems the water was caught unawares as it gently rolled to shore as it always does. Beyond the ice, whitecaps wink on the open water. No birds out there, no gulls along the shore. I head back to the house, fingers numb.

When I arrive, Brian tells me he was down by the wetlands and saw a great blue heron, huddling behind some rushes. "It looked thin," he said, unwinding his scarf.

He sits down at his desk while I think about why the heron was there and when the egrets will return to stir up the muck with their gold feet.

Stung

A LATE AUGUST NIGHT, sultry and humid. I swept crumbs off the counter, shoved the last dishes into the dishwasher, scrubbed the grease off a hefty spatula for the grill, and pushed open the screen door. Something the size of a hummingbird caught my eye and darted away, but of course it wasn't a hummingbird at that hour and in an instant my husband standing nearby, just inside, said he'd been stung. We heard intense buzzing and saw a massive hornet whirling furiously around the kitchen ceiling while others were pinging and thumping against the screen. Brian grabbed a fly swatter and flailed around the kitchen, waving at the hornet in midair, leaping at it and sending it spiraling upward, so my daughter seized the swatter and we waited, paralyzed. Quiet for an instant. It had vanished from sight on top of the kitchen cabinets in a space about two feet from the ceiling—we thought. Sophie moved a chair, climbed stealthily onto the counter, peeked over the top of the cabinets, and nodded—she held the swatter up about a foot from the target, aimed, suspended, wham. Got it, she pronounced. We breathed. Brian said his toe was on fire.

We knew that hornets had a nest under a vent outside the kitchen door. We'd seen one about a month earlier on a screen at the back of the house and had gingerly brushed it away, stunned at the breadth of its abdomen and length of its nimble legs. My daughter and her two daughters, four and not yet two, were living with us and I'd spent months instilling in them a respect for everything alive. (We scooped up spiders on a bedroom wall and tossed them outside; we palmed worms and slugs and returned them to the dirt.) We figured the hornets would keep to themselves.

By midnight, Brian had taken Extra Strength Tylenol and I was bringing one ice pack after another to stave off fiery needles stabbing through his toe. Tylenol clearly did nothing, so at 3 a.m. I handed him some hydrocodone, prescribed after some nasty gum work and saved for another occasion. We should've thought about allergic reactions (bronchospasm, cardiac arrest, anaphylactic shock all possible) but we just wanted to ease the pain.

At 8:00 the next morning I was on hold with a pest control company. Brian, somewhat delicate, was limping toward the garage for a can of Raid.

"Are you kidding? You're not going anywhere near that nest."

"It's no problem, I'll just get a ladder and spray like crazy."

"No way."

The next day a guy appeared in full bee regalia, a white suit covering every millimeter of skin and a wide-brimmed hat draped in gauze. He sprayed powder under the vent and scraped out the nest, which he said was home to about a hundred European hornets. We used another door that day, keeping the kids away and telling them the doorknob was broken. As the poison took effect, bodies dropped softly into the grass, scores of them, amber cusps that caught the sun, hundreds of curled black legs. For the next few days, pest control told us, foragers might return to the nest and get upset that their home had been wrecked, their colony destroyed. They might behave erratically. I would. We were advised to stay away and to clean up the dead tucked in the grass—which could still sting.

Before Ella was up in the morning, I was out the door with a dustpan and brush to sweep up the bodies. Brian wondered what pollination would fail to happen and felt bad about the slaughter. I didn't care, I truly didn't. Blindsided by a hornet, I'd gone into protective overdrive, thinking only of the girls. Overarching respect for living things had screeched to a halt and reversed: kill the hornets. I was a predator, a hawk veering down at a mouse—or was I potential prey, threatened, a mouse terrified of the hawk, all systems askew?

A few days later, Ella calmly asked, "Are the wasps all gone?" I still don't know what she knew or how. "They found a new home," I replied.

<hr />

My daughter soon took her girls back to Brooklyn, and the house was tangibly quiet until I walked into the living room one day and heard furious buzzing overhead. On the skylight about twenty-five feet up, a European hornet thwacked and spun, struck glass and veered off, looking for a way to exit. I opened the back door and foolishly hurled a few pillows in the hornet's direction to usher it out, but never came close. Moving fast, we carried in a tall extension ladder and rested it precariously on a beam that spans the length of the room. Brian climbed, Raid in hand, I purported to hold the ladder steady. I opened all windows, fearing carcinogens with every breath. Wavering slightly, he aimed and sprayed. Almost instantly the hornet's fury subsided. It crawled achingly along the sill of the skylight, this homeless forager. Every few moments, it dropped a few inches, but managed to fly back to its foothold. We didn't want to spray any more, we had to wait. The scene felt pathetic, the hornet was pathetic. Back and forth it inched, stumbling, falling, climbing. I did chores in the kitchen and returned to the doorway every few minutes to watch. Just die. At last I heard a faint tap on the floorboards and rushed over. The hornet was limping agonizingly, almost sideways toward the open screen door, dragging its thin double wings. I grabbed a magazine from the coffee table and ended its struggle, and mine.

Virtually the same scene recurred two days later. That time the hornet's battle at the skylight was more vigorous. That time I had to go out briefly and by the time I returned, it was gone from the sill. I searched the living room floor, peeked under chairs and behind sofa pillows, but I never found it. The screen door was open, the sky sapphire blue, sun hot on the deck, and I wondered if it had somehow wobbled over the threshold and made it outside to die. I was coming close to sentimentalizing a death that had taken too long, longer than the other, but it would've been dishonest. Still, the impetus to escape its accidental prison and return to sunlight struck a chord. What will, drive, lack of doubt, sense of belonging, need.

European hornets don't pollinate, or do so only accidentally after licking nectar from shallow flowers. They scrape the bark off birches, lilac, and rhododendron, suck the sap inside, and use the wood for nests.

Do they do anything we call good? I wondered if Brian would now feel justified in eviscerating the nest, and should he? Was our action *good?* Hornets eat honeybees and yellowjackets that pollinate, another strike against them, but they also kill flies that bite us and land on our sandwiches with bacteria-laden legs. Ethical valuations are quick to falter.

The shush of waves on a beach makes us feel good, and seems inherently good, as does the sight of daffodil spears nudging through the dirt in March, but aesthetics drives this morality and others. Death has its own arena, often amoral. An osprey snatches a vole because it benefits. We slam a hornet because the family benefits. We raise farm animals for slaughter in spaces too small for them to turn around in because it benefits our pocket. Is morality now a factor because this method is cruel—any more cruel than a hornet that lays its eggs inside a living spider—or because we have a choice to do things differently?

Most people would not kill the animals they routinely eat. I don't know if I'd cut the head off a turkey at Thanksgiving. Like the titular character in Herman Melville's "Bartleby the Scrivener," I prefer not to.

Now when I go out the kitchen door, I glance up at the vent, bright aluminum shining in the morning sun, no vestiges of warfare on the shingles or under the eave. I remember one evening in the '80s, trudging up five flights to my apartment on the Upper West Side only to find my three locks had been gouged off the door, leaving a gaping hole. Sweaters and dresses were strewn across the floor, stockings dangled from open drawers, the plastic buttons on my mattress lay exposed and maddeningly silent. Someone's hands had fingered through my underwear, checked the coffee tins in the refrigerator, pulled out my books, rifled through my notebooks, and peered inside my canisters of flour and sugar. Someone now had my grandmother's gold bracelet and wedding ring. The black and white TV, still there, laughed in my face. Shaking, I called my boyfriend and spent the night at his place, unable to rationalize the chances of a second break-in on the same day. The next night, I shrugged and moved back in.

Hornets don't reuse an old nest. Had they survived, they wouldn't have come back the next spring. One haphazard hornet drawn to the kitchen light or to the smell of my tomatoes screwed it for the rest. It's all too easy to think of the incident that way.

The house felt empty without the kids, so I walked a quarter mile to a beach on Peconic Bay where we had gone together all summer. On the double yellow line lay a smashed turtle just a few inches in diameter. The flattened carapace with tiny orange strokes and circles unique to this one turtle opened an unwarranted sinkhole of feelings, likely because helplessness and sorrow and outrage stack up daily just by reading the news: apocalyptic flooding in India that swept away families like tree limbs; the ivory billed woodpecker with its bright red crown now "probably extinct"; a massive, triangular ice chunk splitting off Antarctica, the break line as clear in aerial views as the crack an x-ray revealed in a vertebra in my neck years ago from an exuberant dive. The arrowhead piece of bone held on, with the potential ever present to sever my spinal cord. With the ice, who knew how long, who knew the effects of life on ice and beneath it.

Once at the beach, I headed for a narrow stream that spills from a lagoon where I'd seen a family of swans the previous day. The regal parents had lifted themselves from the surface now and then, stretching and flickering their wings, bright white and water dappled in the early sun. Three gray adolescents, adept at foraging on their own, glided in the shallows and ducked their heads, searching where they liked. Today plovers skittered along the shoreline and the water was flat as tin foil. Ahead I noticed an odd globe of gray in the stream. As I approached, I knew—even though the young swan's head and neck must be concealed beneath the body. How had this happened, and why?

Evening tides carried the body into the lagoon where I found it the next day floating like a plastic bag of trash. A gray swan that I took to be its sibling was feeding at the mouth of the stream, and I watched for a long time to see if it sensed the dead one and would go to it, but after a few steps in that direction, it turned, sank onto the water, and paddled off, glancing slightly side to side for minnows.

FIRE

Eclipsed

M Y HUSBAND SAYS THAT India has built one of the world's largest solar farms, big enough to power 1.3 million homes ... and China will likely be selling only electric cars by 2035...and soon we'll have blood tests that reveal hard-to-detect cancers. Such facts slide off his tongue over dinner on a Monday night. He does the *New York Times* crossword religiously and rapidly, Monday through Thursday. He knows things about stuff. On December 10 Brian confided (with a tentative glance in my direction), "The moon hasn't done this since 1638. I'm setting an alarm."

"Good idea," I chirped, having no intention of getting up, nor any palpable grasp on my failure of imagination. I entered my daughter's room and gently closed the door, as if stealth would negate my betrayal without cause. Linen curtains tied erratically to a rod quivered over the radiator. Behind cracked glass was a huge poster that used to be mine, a black and white shot of Jimi Hendrix taken by Linda McCartney. A pink silk elephant from Thailand graced the windowsill along with a bunny and teddy

bear slumping with age. In this room just big enough for a bureau and a
bed, I would hear no alarm, nor his stertorous breathing—the seizing of
air that awakens me at any hour, being so deliberate, so mortal, punctuated
by fluttery exhalations whispering quiet defeats of which I'm unaware.

Earlier that night I stood alone in the backyard watching a fat moon
rise over the peaks of my neighbor's house. The surface swelled and shone
like mother-of-pearl skin stretched taut. I remembered my midwife who
claimed the maternity ward always overflowed on nights of a full moon.
Gravitational pull, she had said. My daughter had stayed high in the uterus
showing no signs of making an entrance long past the time when the
medical world said she should. Two weeks late, two extra weeks growing
plump in the dark, dreaming dreams without context, she descended when
the moon waxed full, winking at her birth, two hours before a scheduled
caesarian.

I woke up sweating. The bedside light was on, my computer open on my
lap. I still wore a hooded sweatshirt over a thin shirt, damp across the chest.
With a colossal force of will I clunked the computer onto the floor, swept
away the covers from my legs and yanked at the sweatshirt. 3:23 glowed
from the pink alarm clock. Why undress? In a semi-dream I realized with
greater acuity than before that the moon was doing something strange or
the earth was doing something to the moon that it hadn't in four centuries and I should witness what that was because it wouldn't happen again
for another eighty or so years which might as well be four hundred. So I
pulled on some jeans and padded past the gaping dark of our room, from
which long low soughing arose from a swaddled form. He must have seen
it, I thought. He had a plan, and he is a man who follows through. Down
the stairs I went, not even waking the dog, threw on a parka and hat and
boots and unlocked the front door and stepped down the bluestone steps
we'd laid a few years ago when Brian decided to renovate.

The moon had freed itself from bristling oaks, crossed the sky, and
now posed blood orange over the Hudson. Straight black limbs stretched

toward it, twitching slightly. In the hours I had inadvertently slept, it had shrunk, aged, receded. Shadows like wet stones mottled one side, but the other gleamed, a dazzling, painful red. Moon without sun. Sinister and spectacular and alone. I should wake him, he should see this, maybe he has, I can't talk about this. Eye wide open, as if the night would consume it alive—deep red eye burning into mine, radiating a splash of needles from the pupil at its core. Nearly lightless, nearly deathless. Stone and crater, crater and stone. I closed my fists in my pockets, tensed my thighs, breathed sharp scallops of air, and tried to swallow four hundred years in a gulp. The moon needed no reason, and I had no explanation.

The earth, that was it, the earth's particles, Brian had explained. I would ask him in the morning if I let on that I had seen it. But he must have... I nearly believed. And if not, what vision would he have missed?

Wrapping the star-pricked sky around me, stamping the auburn eye in my mind with the conviction of never forgetting—as if one could remember such things as your baby's first sound or the remote look in her eyes under neon light and bodies looming like planets above, big and amorphous and unexplained. Would she see the moon smoldering red in eighty years? Or would her daughters? Or would they, missing her, wish she had seen such a sight? Those shadows spreading like smoke over the lovely orange face of the moon were absurdly sad and heroic. Especially on this day that had to happen for this effect to occur, this day of least light of the year in the northern hemisphere when people used to entreat the sun to return and now simply assume it will. I thought of my mother sleeping deeply in her hospital room, on her back, her new floral nightgown open at the neck in the overheated room where she sits all day, nodding involuntarily and sleeping while light dreams tumble in her mind, letting her believe her mother will pick her up and take her home the next day, but each waking day grows shorter, and she worries that she has not packed. I thought of the confluence of eyes on this climactic moon and of cloud-covered countries that would see computer images after the fact and be satisfied, and of millions sleeping unaware that for a meticulously recorded few hours the moon was struggling to glow like graffiti in a subway tunnel. Was my standing in the street in the heart of the night a lie? Without him I would not be there. But to wake him, to kiss him, would break the spell. Alternatively, as the fairy tale goes, the spoiled princess throws her frog against a wall in disgust, or lets him sleep on her pillow all night. Either way, through violence or sex, sex and violence, presto, she has her prince. She didn't even have to believe.

Dark blotches backlit by orange were encroaching on the moon's blushing face as I stood there, guilty and alive, not knowing how long was long enough and not knowing much more than those who saw such a moon in the centuries preceding 1638 and not knowing if omission constitutes betrayal any less than does an eclipse of the one you love.

Where was the sun-lit moon hanging over a mountain in Wyoming that I saw when I was fifteen, alone and virginal, two thousand miles from home, sitting on a sandstone ledge and squinting at the silver light (or at the fact) that a man in a high-tech suit was bounding over her surface insensible to her gravity, reporting data to earth, and implanting an immortal flag? That hour was well before the shared life—when unrecorded nights with someone were the stuff of dreams, and you believed that someone, like a shadow betraying the missing part of a quarter moon, was needed to complete you.

Everyone I knew (and a boy in New Hampshire I didn't yet know) watched Armstrong take the historic step in black and white, marveling as much at the fact that they could watch from a couch in the den as at the completion of the mission that Kennedy had promised "before this decade is out." The moment was thrillingly irreversible and American. But something quivered down my spine and tugged like a plumb bob dropped too fast. On my perch above a field and gleaming stream, enveloped in the scent of cedars and dry rock cooled quickly after sunset, loss seeped like blood from a rite of passage whose meaning I could not understand.

For forty years I scarcely thought of that night, recalled it in flickers and starts, maybe, behind wars in Vietnam and Iraq, behind love and childbirth, which eclipsed it.

Brazen moon! Bit of slow burning coal in the sky tonight!

My legs led me up the stone steps, and my jacket fell on a kitchen chair. Without a light I went softly upstairs, re-passed my bedroom door too cold to wake him and explain, heard only my distant breathing as I closed the door on unspoken moments. Sliding between sheets, I pulled the morning-blue comforter over my shoulders and hugged my daughter's floppy bear. Had I not slept I would have heard the alarm an hour later but no footsteps and no door.

If the earth were not cloaked in air, the moon would have merely turned black in its shadow and re-emerged unblemished and white, with no breathing being to witness it. But the dance of particles around the earth that I couldn't see assumed a stage thousands of miles away in the amber light of the moon, deepening umber and angry red. I awoke by chance at the moment of complete eclipse. He set the alarm twice, got up twice, peered through the bedroom window to see a gray shadow beginning to inch over white and hours later the same image in reverse, too early, too late.

Speaking Light

A HALF-MOON RISES OFF the coast of Hawaii and stars begin to appear scattershot across the sky. In the shallows two tiny tentacles part the sand and two bulbous eyes emerge. A polka-dot body and eight arms quickly follow, floating through tenebrous water. A bobtail squid, about the size of your thumb, is looking for shrimp and crustaceans. At the heart of its survival is a language of light—a calculation of light in the sky that singular night and its own measured creation of light. The squid casts no shadow; it shimmers in the water among the reflected, moving moon and stars.

This squid was born in water teeming with billions of bacteria. With its first breath, the flutter of gills began stirring up a swamp of bacteria and attracting the cells to its body. At the same time, the squid's mantle cavity in the core of its body emitted mucus that trapped the cells though most would remain outside, useless, and rejected. Highly discriminating, the bobtail allowed only one species, *V. fischeri*, into its interior—inside a trinity of pores, down ducts, and into a secluded crypt in the mantle cavity, a journey that seems as vital as that of an egg from ovary to uterus, as sacred as worship in ancient caves. Here the bacteria multiplied and shed the delicate flagella that had propelled them. Mucus coated the interior, protecting the bacteria, even as the bacteria themselves instructed the exterior tissues to regress and die off, effectively sealing off admission. Now the little squid, just eight hours old, began a lifelong symbiosis with *V. fischeri* in its cavity, the two together making light.

Everything moved fast, all cycles expedited in a brief life of two or three months. Every day during daylight hours the squid ejected about three quarters of the symbiont bacteria into the water, giving space for the remaining ones to grow. By evening each day, when the squid emerged from the sand to hunt, its body was ready to glow—at will—the squid providing sugars and amino acids, the bacteria creating light. To start, the bacteria signaled one another to be sure there were enough of them to work effectively. Then a *V. fischeri* enzyme called luciferase combined with oxygen and luciferin molecules causing a chemical reaction—and behold, light.

When the moon waned and clouds drifted over the stars, the squid tempered its light by limiting the oxygen it supplied to the bacteria. More nuanced still, the tissues of the mantle cavity acted as a retina modifying the amount, not of incoming light as with us, but of outgoing light. A seal swam below and glanced up and missed the squid with its dim shifting gleam. A lizardfish passing above failed to see the squid because the squid's ink sac concealed its light organ from above. Reflective diverticula in the sac directed light downward, contracting like an iris to cool luminescence on that dark night.

The ability to generate light evolved independently many times in our oceans, making it the most common language on the planet. Crustaceans, echinoderms, gastropods, coelenterates, some sharks, and at least 1,500 species of fish create cold light, which glows without incandescence. Likely there are more since much of the ocean is deep and utterly dark, necessitating ways to navigate. And speak. A jellyfish flashes exclamatory light to startle a predator; a siphonophore produces a veil of light, unpunctuated smooth talk, to mimic plankton and throw off a predator; a brittle star sacrifices the brightly lit tips of its arms to distract a predator while it scrambles away for safety; a female syllid fireworm rises from the sea floor under a full moon and glows to attract a male. The tactics are diverse, the language supple, abstract.

The color of this language is usually blue-green because short wavelengths of light travel well in water while longer ones of red and orange do not. But in the depths of the North and West Atlantic, a skinny fish with oversized jaws emits red light and sees it. Glowing red, the dragonfish becomes invisible to nearly all other creatures. It sneaks up on prey and grabs it with long hooked teeth. It makes itself alluring to another dragonfish, who alone can see it. Red is food, red is sex.

The slate-blue, deep-ocean eyes of my daughter's newborn find light filtering through a dusty skylight, or yellow light beneath a shade across the room. They stare for long minutes, meditative, receptive, and unbiased since there is no context. Is this a binary language of light and nonlight or something more nuanced? The sheen of a lightbulb or of the moon is as abstract as music. Dull suns loom above her body and move on as we carry her here and there, her pupils black holes in the rings of her irises, afloat in milky white, small planets in a new universe, pulled toward light.

At one day old, her eyes seemed to say she had seen all this before, that she knew far more than I. Her gaze hinted at something ineffable—quickly snuffed as her eyes rolled back, lids trembled and closed—leaving me to question when and how the purely physical transcends its own being.

⚌

The painter Wassily Kandinsky worked for decades to elicit the spiritual through color and forms that transcend anything we see in the phenomenal world. He shifted the vocabulary of painting away from realism and beyond impressionistic or cubist depictions of the recognizable— Van Gogh's haystacks, Picasso's guitars, Matisse's tables and flowers—no matter the distortion, no matter how relative to the artist's perception. He wanted art to approximate the abstraction of music, which seems to me analogous to a newborn's abstraction of unnamed, unidentified light.

His *Composition 8* is symphonic at a glance, without the duration of music. Triangles crescendo, half circles bubble along like a basso continuo, a rosy orb with black and purple interior rings pulses in one corner, a red circle sirens as it invades or escapes that bigger sphere; lines run parallel and askew, clash and cross, a squiggle whispers through a line, suns sizzle and glow, lines, squares and circles collide demolishing spatial perspective as if all orchestral instruments sounded at once without melody or finitude or order.

Kandinsky painted sound and heard color. "Light blue is like a flute," he wrote, "darker blue a cello; a still darker a thunderous double bass...." When he looked at yellow he heard the sound of trumpets "strong, harsh, and ringing." Color, he thought, also had inherent movement. Yellow is ex-centric, spinning out from its center, aggressively extraverted, while concentric blue curls around itself like a dog in a cold shed, looks inward as into a well, holds the possibility of depth. One can see pure color only when unbounded in the mind "which exercises a definite and an indefinite impression on the soul." Words, recognizable form, names for and associations with those forms—all interfere. They're reductive and inadequate. They don't yet exist for the newborn gazing at light.

Synesthesia like Kandinsky's opens channels of expression, leads me to hear a sheet of light over water as a wash of cymbals or see the warning of gulls as dashes of lightning, which may or may not sharpen the acuity of either sense. I have also translated the bobtail squid's bioluminescence to spoken or written words to explain its function, a metamorphosis that is never accurate. But not always reductive, as Kandinsky would claim. I am not less impressed by the complexity of the process, the beauty of two beings together making light, the nuance of expression of which the squid is capable. I only feel more compelled to lean on language, expose its limits in mimicking the physical world and translating its meanings. The planet is talking; we can eavesdrop, knowing one always misses key words.

When I was eight I ran into the Caribbean and lit a rippling field of silver-green across the surface. The water was on fire. I halted and it turned to ash. With a splash it glowed again, magical cold light. My father smiled and explained, but I didn't listen. I thought the stars had fallen, or another galaxy of stars lay underfoot that suddenly I was privy to, but why me and why then? I thought that planets and moons bobbed about me with life on them, electric energy in secret cells.

Sunshine

M Y LEFT ARM SHOOTS up, ball lifts from my hand continuing
the rise, a fluorescent-yellow globe slips into the globe of the sun,
a perfect fit, for less than a second before my racket strikes. A blinding
afterimage skims across the net and around my opponent—red orbs, fiery
and chaotic, more like the sun at its core than the gold image my ball
briefly eclipsed.

Sunlight splays across the street, laying down shadows of straight pines
and gnarled oaks, chimneys, rooftops, all reduced to two-dimensional
pattern and structure. The sun frames our day in ways less obvious than the
coming and going of light, which is noticeable from my kitchen window
when a lamp brightens the gloaming in a house nearby, and poignant in
the summer Arctic when the sky turns dusky at midnight and one wonders
if it's time to sleep. With the absence or presence of sunshine, midsummer
leaves droop like bat wings or shimmer like pinwheels; hours of the day
unroll monochromatically or seethe with possibility. The sun becomes
what it creates, in us or for us. A picnic in the park; a pageant of orange

and magenta and green that lingers, sun reminiscent, after the horizon has swallowed the sun and charcoal clouds dissolve.

———

The light leaving the sun's surface takes eight minutes before it reaches the rock on which basks a snake, relishing the warmth of the rays. But light traveling from the core to the surface may have been born 30,000 years before in a dense, sweltering place beyond anyone's conception of hell. In the center of the sun, 150 times denser than water with 200 billion times the atmospheric pressure of earth, hydrogen atoms smash together, four of them squeezing together to produce an atom of helium and emitting a fraction of that matter as energy, which cumulatively swells to a staggering flow. Neutrino particles stream out constantly. Light particles bounce back as if hitting billions upon billions of mirrors and escaping only after tens of thousands of years. Propelled from the inner core, they travel by radiation to a boiling convection layer of whirling gas in the outer core, where they hitch a ride on thermal columns shooting out to the photosphere, the visible surface of the sun, the yellow sun we see, which is not the actual surface of the sun.

We know the sun by its effects and illusory appearance; outer layers—the burning red rim of the chromosphere and white plumes of the corona—are visible only during total eclipses of the sun, a phenomenon that historically evoked fear and possibly caused heads to roll. Neolithic petroglyphs at Loughcrew in Ireland show an alignment of sun, moon, and earth, which NASA believes depicts an eclipse in 3340 BCE, the earliest on record. Nearby lies a basin with the remains of forty-eight people who may have been sacrificed to save the god of the sun from dying. In 2134 BCE Chinese Emperor Chung K'Ang allegedly found that his two astronomers had failed to predict that a dragon would devour the sun. Irate, he had them beheaded. A Greek poet, Archilochus, fretted over the eerie loss of the sun in 647 BCE and puzzled over the intentions of the gods: "There is nothing beyond hope, nothing that can be sworn impossible, nothing wonderful, since Zeus, father of the Olympians, made night from midday, hiding the light of the shining Sun, and sore fear came upon men." Anything was possible. Archilochus was wise, not cowardly, in realizing the potential for disaster and inherent power in the extraordinary.

Today, with greater hubris, we greet these phenomena with excitement. The media pumped up the 2024 solar eclipse weeks in advance, everyone bought special glasses, and devotees traveled long distances to catch the eclipse's swath from Texas diagonally north to Maine and on into Canada. One woman flew from central France to watch the event over Niagara Falls; a man in Amsterdam flew to Austin where he caught a flight specially scheduled to cross the path of the eclipse over the Missouri-Arkansas border. Many described feelings of awe and reverence. From the south shore of Long Island, I watched the moon slip serenely over the sun and blithely take its leave, while the sun, reduced to a moonlike orange cusp, recouped its position and power.

Photos depicting the elegant progress of the eclipse are strikingly different from those of the photosphere at close range. Images taken by the Daniel K. Inouye Solar Telescope in Hawaii in 2020 show it looking like peanut brittle—a lumpy map of cell-like structures, which are hot pockets of excited, rising, electrically charged gas called plasma. Bordering the cells are dark lines where the gas is sinking and cooling. Temperatures range from 7,000-11,000 degrees F, an oddly cool interlude between the raging core at 27 million degrees F and the corona at 3.5 million F. Relatively small explosions of the plasma, called nanoflares, happen frequently on the sun's surface. Although they dissipate quickly, these staccato eruptions might be the source of the corona's intense heat, which has puzzled scientists for many years. As a layperson gazing at the sun, expecting it on the east side of the house in the morning where I sip a cup of coffee, the dissonance between what I see and what is occurring leaves me uneasy, as if I were one of Plato's prisoners giving more credibility to the shadows on a wall than objects moving in full light outside of the cave.

All is motion. This seems true, is true. When the sun, clearly delineated as a copper penny, sinks below a hill, I can see how fast the earth is rotating and wonder why I don't fly off, like a kid slipping from a rapid merry-go-round. I don't think about the fact that the sun, though not a solid, is also spinning—slightly faster at the equator than at the poles—releasing solar winds that shoot out radially through the solar system, carrying a magnetic field that creates a sort of bubble in which planets and stars and asteroids and comets float as in a massive caul. But some of the ionized gas in the solar winds keeps rotating, bound by gravitational pull even 20 million miles from the sun, as if held in colossal arms, an invisible encircling embrace. Scientists don't know precisely why or when solar winds shift from rotational to radial. Weirder still, some of those rays of energy in magnetic fields shooting directly at us (the way a kid draws the sun with spokes) suddenly flip back, reverse direction entirely, and aim back at the heart of the sun. In the end, solar winds carrying energy are whipping out and around and back in chaotic paths, even as I glance at a pale pink dawn or gaze at an orange orb at dusk, even as I can read to the minute when the sun will rise and set.

When I was twelve my parents took me and my best friend to Jamaica for a week in March. We ate succulent pineapple and papaya and flirted with boys on the beach while my mother read paperbacks and tanned her legs. The day before we left, she eyed me over her glasses and advised me to work on my tan. I didn't even look as though I'd been to the Caribbean, she remarked, as if it mattered. I smeared on some baby oil and lay by the sun-lacquered water, feeling a faint breeze on my arms, and deep heat flowing through my whole being and into the plush towel on which I dozed with red spheres blooming behind my eyelids.

The next day, several hours into our flight, I complained to my mother that the dress I wore was irritating my chest. She told me maybe the cotton was a bit itchy. There wasn't much I could do as I had nothing to change into. When I got home I lifted the dress above my head and found that my entire chest from breasts to shoulders was a massive yellow blister. We put on some cream and bandages, and that was that. For days I had to sleep on my back.

Decades later I sat by a lake with my two small children, my mother on a towel nearby. She eyed my leg and pointed to a mole that I'd barely noticed—like any new mom, I prioritized the bodies of my children and figured mine was fine. It looked a bit smudged. You should have that checked out, she advised. The smudge turned out to be a level-two melanoma, which had to be excised more than once.

Now my copper-haired granddaughter scampers around on a beach with long sleeves, her face and legs coated with sunscreen. She yells every time we apply it and struggles free. Sunburn will hurt, my daughter coaxes. You need a hat, it's sunny. She is growing up with constant reminders of danger even as we exclaim, Such a beautiful sunny day, let's go out!

———

The sizzling corona, the earth aflame, ruby circles behind my eyelids after throwing a ball at the sun—these are haunting images detached from a moment of perception, bent out of time as gravity bends light. The sensation of perception or traces of it flash in the mind's eye the way a hummingbird arrives and flees before the mind reconstructs its whir and buzz and a splinter of mint green. While the seemingness of the sun is present, here and now, we are always at least eight minutes behind, an eternity in the flight of a hummingbird or the dart of impulses along neurotransmitters in the brain to elicit a response, even a thought.

There's an odd symmetry in the emergence of light from the core of the sun to its surface and the absorption of light by our eyes and its journey to our core, the brain. Light passes through the cornea, iris, and gooey vitreous humor at the center of the eye. At the back it hits the retina where millions of nerves cells convert light to electrical impulses and send them along the optic nerve to the brain, which has to receive information from both eyes and connect the two before converting energy to image.

All is motion; all is metamorphosis: from hydrogen and helium to energy to its absorption and decoding in our fallible minds as we witness the play of light on a penny, water, waves, a windy meadow. And only then translate, as Impressionist painters did, the dance of molecules in a color spectrum into a *sensation* of being by the water, a meadow, pinning down what essentially cannot be pinned down.

Folktales and legends from around the world try to explain our vital relation to the sun through binaries of light and dark, god and mortal, male and female, life and death, salvation and sacrifice. Many portray a god as originator of life, and therefore of the sun, created purposefully and deliberately. Many paint men and women as active participants. One Indigenous folktale, told with variations among different tribes, paints a different sort of scene, a time when birds and animals flourished on the earth but humans had not yet appeared.

Rabbit is ambling along one day and decides that he can't see well enough to find his way. He decides there isn't enough light and says so. Owl disagrees and says he wants more night. The two argue for a while and then, reasonably, call a council. Raccoon and Bear side with Owl. Even Frog says he can't sing well in too much light. But many birds agree with

Rabbit, needing more light to find sticks to build their nests, while Buffalo wants light to search out grasslands to graze. The two sides cannot agree.

Rabbit and Owl decide that whoever has greater power shall be the one that decides. They test this power with words, one saying light, light, light, the other, night, night, night. If Rabbit errs and says night, he will lose; if Owl slips up and says light, he will lose. Rabbit picks up the tempo, pitching light, light, light. Owl follows the pace. Faster and faster they go. Back and forth until Owl, saying night, night, night, hears someone say light, and says night, night, night, light. He quickly realizes his error. Too late.

Rabbit carries the day, and for a while there is more light. But then he realizes that some animals need the dark for sleeping and hunting, and so he compromises, allowing time for night.

While poet Archilochus put the existence of light and dark in the hands of a god sitting atop Mt. Olympus, this story gives the power to animals whose eating and sleeping is inexorably tied to their surroundings. It lends power to language and its capacity for creation. It demonstrates the relative nature of perception—for Rabbit there isn't enough light, for Owl, too much—and yet, they compromise.

As dynamic as the sun is, it is middle-aged, about halfway through a projected lifespan of 10 billion years. One way to follow its course is through a counting system developed in 1755, which is based on cyclical shifts in magnetic polarity. Solar Cycle 25 began calmly in 2019, but activity is rising and will reach a crescendo in 2025.

Every eleven years the polarity of the magnetic field reverses; during that period the magnetic polarity weakens to zero and then rebounds in the opposite direction. Smooth magnetic field lines on the sun get tangled and sometimes burst, erupting in volcanic solar flares (a billion times the size of a nanoflare) spitting out radiation at the speed of light. Fiery tongues blasting earth can disrupt satellites and knock out power grids. A solar flare in 2021 caused a radio blackout over the Pacific Ocean. Some right-wing politicians, more obdurate than Owl and Rabbit, blame solar flares for climate change. At a meeting in 2021 one Texas representative inquired, "We know there's been solar flare activity, and so is there anything that the National Forest Service or BLM can do to change the course of the moon's orbit or the Earth's orbit around the sun?" There followed a moment of silence.

The tangled magnetic lines can also cause coronal mass ejections (CMEs), which are colossal explosions of plasma with embedded magnetic fields. Slower than solar flares, they take anywhere from fifteen hours to several days to reach us where they, too, can damage power grids and even disrupt your GPS. If an astronaut were walking on the moon, a CME shock wave hitting her would deliver radiation equivalent to 300,000 simultaneous chest x-rays. It only takes 45,000 to kill you.

The sun's pulsing tangerine surface with flashes of gold sparking and dying seen in NASA images strike me as images in a crystal ball where the earth itself incinerates, all life and water scorched, eons before the sun consumes us, as it may well do. With greater conviction, scientists say the sun will swallow Venus and Mercury. Mars might evade this fiery end, left intact with its ice caps and toxic water sealed in stones. On its red dust manufacturing plants might still stand (those long dreamed of by billionaire techno-wizards) humming and hissing without purpose.

Someday our star will swell in a spectacular and grotesque crescendo as hydrogen fuel runs out and the sun's mass fuses helium into carbon. The energy produced will increase and rush outwards, inflating the sun up to 200 times its current size. But because the gases are distributed over a greater surface area, that area actually cools somewhat and turns red. The expanding red giant will lose mass as solar winds sheer off outer layers. Gravity, the force that has held us, kept us from getting lost, will weaken.

What then? It's as if we were kids asking, if you had to die which would you rather do, burn or freeze? Vanish in a blast of fire or disintegrate cell by cell in a lonely, lightless place? Were we just a tad farther from the sun, say some scientists, we might escape the fate of Venus. But as chaotic as the life of the sun is, so are the unknowns circling its demise. Planetary orbits might shift by the time the sun becomes a red giant, ultimately a white dwarf, too small to be seen by the naked eye. The earth might spin off like a random stone a child threw in a pond, yet there would be no bottom, no resting place, or like a baby sea turtle that never laid eyes on its mother and struggles to the water where totally alone it circuits the ocean for years because its genetic code has said it should, yet there would be no code, no resting place.

Because we've given the sun a jump start on frying our planet, and because our mother star will shrink to a ball of captured electrons and flicker out, I think about what other suns could light ocean floors and ravines and what eyes could gaze into chaos without being burned.

Fiery Dragon

CLACK! CLACK! THE END of a plastic tray in the maternity ward tumbled to the linoleum below. The perpetrator lay on the tray, several minutes old, two nurses at her side poised to take her vital signs when boom, a leg shot out and hit the end piece with fury, or indignation, or unrecorded *joie de vivre*. Ella was average height and weighed seven pounds. Seven pounds of will and spirit. She had a blush of red hair though both parents are dark brunettes. The red gene had gone underground for some generations, all the way back to Ella's great-great-grandmother, and had burned there like a coal fire—waiting—as scores of progeny with walnut locks quietly went to school and worked like Puritans to eke out quiet lives. This can't be, we said, eyeing Ella's hair in candlelight or dusk when the copper cooled a bit and we felt sure would simmer to ecru. But her wispy orange eyebrows told a different story.

In Centralia, Pennsylvania, a vast underground network of coal mines has been burning since 1962, as deep as 300 feet and expanding steadily in all directions. The fire spews sulfurous gasses, chars surface trees and plants, and creates twenty-foot pits hissing fumes. Ventilation in coal mines supplies the oxygen needed for coal to burn. This one allegedly started when some guys burned trash near the entrance to the mines. Sometimes such fires start spontaneously: if coal is near the earth's surface due to erosion, temperatures of about 105 degrees can ignite it. A streak of lightning may set it afire. Some coal fires survive undetected for decades, burning downward hundreds of feet, consuming oxygen between bits of dirt or in fissures in rock. The one in Centralia, which has turned the place into a ghost town, will persist for a few hundred years and destroy about 4,000 acres before the energy supply runs out.

———

At three months Ella raged against sleep as against the dying of the light. She raged when she was hungry, raged when she was bored. I'd carry her a mile in a Björn, singing calypso songs about cheating lovers, before she'd let go of her waking life and succumb to an abyss where anonymous forms floated soundlessly, amorphously. At twenty months she exploded if her bike hit a rut or her banana broke in two. She'd stand in the kitchen in her lemon pajamas demanding I peel her another while I—silently fuming at the futility of reason, withering from blitzkrieg demands for jam in the yogurt and a different cup (*no, no, no not that one*), and hankering only for a sip of coffee and a glance at the *Times*—yes, I capitulated on more than one occasion.

———

Firefighters considered dumping water on Centralia's coal fire, but they couldn't inundate an area so large, and they would have had to keep pumping for several years (far longer than the floods God visited on his world). They dug trenches as a firebreak, but the fire leapt ahead of them and the trenches only served to aerate the coal. They bored holes around the coal fire and filled them with gravel, sand, and cement to block oxygen, but the searing heat of 1,000 degrees F relentlessly ate away at the fill, leaving pockets of energizing air.

On record, Ella has screamed for forty-five minutes straight, a continuous conflagration. This occurred mysteriously on her waking up from a nap at about eighteen months old. Whether it was dreams that haunted her or the confounding transition from that world to a waking one beyond her control, we will never know. Later, the issue was clothes: a dress. Red, her favorite color. Ruffled shoulders, ruffled hemline, too long to play in, too fancy for school. But she would have it, her "summer dress" at every opportunity. She'd pull off her pj's and rip off her diaper at 7 a.m., flinging it on the kitchen floor. *I want my summer dress.* I'd zip up the back and she'd wear it with her red Keds and red baseball cap while eating oatmeal with maple syrup before waging war with her mom over what she would wear to nursery school. Methods of containment largely failed. Reason and logic failed. Pretending not to hear her failed. Bribery might work but was obviously not optimal in the grand scheme of child-rearing. At last, we would pick her up and take her to her room and just sit there, stonily, waiting it out.

Red rock tops many hills in eastern Montana where coal fires have burned for four million years. As coal turns to ash and collapses, it allows air to penetrate deep in the soil where fire persists, heating and fusing and transforming the materials around it and especially above as heat rises. Sandstone turns brick-like, shale to a hard ceramic. Fused rock is called clinker, onomatopoeic for when the stones knock together. Since it's porous and resists erosion, it survives on escarpments and outcrops across the American West, protecting the underlying rock. In Montana, North Dakota, and Wyoming clinker is often called scoria because it looks volcanic.

When her little sister was born, Ella lashed out at her parents, scratching and kicking at unpredictable times. She came into their bed at night and tossed and turned and clambered onto them. She didn't touch the infant, who generally wasn't in her way. But by the time Florence reached five months, Ella was on top of the situation, snatching a toy from her mouth, with an outburst: *she might choke!* And warning her three-year-old peers as they approached the pram in their floral dresses and ponytails, *that's my baby.* Her protective instinct was, at base, territorial.

Even the Arctic is burning, which is as oxymoronic as arming teachers to reduce gunfire. Peat formed from waterlogged, decaying plants in the permafrost may simmer undetected for decades. Peatlands are the densest carbon-rich ecosystem on earth, holding on to carbon, absorbing runoff, and supporting sedges, moss, and heather. Now as snow, which deflects sunlight, melts away and permafrost thaws, the peat is exposed, burning more quickly than ever before and issuing billows of carbon-laden smoke into the sky. Wildfires take off in the spring, not summer. Roughly 1.5

million square miles of peatland with deep stores of carbon lie waiting, increasingly vulnerable, like us.

———

Ella consumes stories with quiet intensity, harboring the words in her memory like reserves of ancient fuel. At twenty-one months she would turn the pages of Margaret Wise Brown's *The Little Island*, reciting verbatim a complex story of discovery and faith, with descriptions of chuckleberries blooming and pears ripening in the sunlight, gulls laying eggs, fog hiding the island in shadow, and storms battering its shores. When I read a book to her that she had only heard once, she corrected me for omitting the word "soon," another time for saying "roared" by mistake instead of "howled." She would snuggle under my arm as we read and focus, hour after hour, scrutinizing pictures, listening, percolating. Once, when we read about a donkey who travels great distances and brings a woman to a barn where she gives birth (our secular reading of the biblical story), she gazed at the last page and kept very still for a long time. There the donkey at last comes inside and looks down at the newborn. Ella, then two, said, "He looks proud." She had absorbed the story as a sapling inhales sunlight, synthesized aspects of the donkey's plight, and drawn an abstract inference from a place I didn't know could already exist.

———

Recently I learned that at its core, the earth has more energy than we humans could possibly need. Most of the super intense heat derives from the arduous birth of the planet, but the supply constantly renews as radioactive substances break down and dense materials move inward causing friction. We think our innermost heart is 5,200 degrees C. Our surface emits 47 terawatts of energy, comparable to the output of thousands of nuclear power plants. The earth is not democratic, issuing more heat in areas where tectonic plates nudge against each other than elsewhere. We call a 25,000-mile U shape surrounding the Pacific, Philippine, Juan de Fuca, Cocos, and Nazca plates, the Ring of Fire, home to the world's most intense earthquakes—Chile in 1960 and 2010, Alaska in 1964—and volcanoes like Krakatoa that rose fuming from the sea.

＝

Rhaaaar, rhaaaar, Ella flung open my bedroom door at 6:16 a.m. *I'm a fiery dragon!*

Don't eat me, I pleaded ducking under the comforter. She roared again with plum fists raised and soft fingers curled like claws. At age three her innate innocence, call it ignorance, has been punctured. She realizes that danger exists. While I tried not to malign the crocodile in a Babar story, she now knows that larger animals may eat smaller ones. She knows a car could hit her, the stove can burn. And somewhere in the semi-real realm of make-believe, she believes a leprechaun could turn her into a toad. Sometimes she demands I be the fiery dragon that sprints helter-skelter in the backyard and catches her from behind as she shrieks, and I (transforming like Saul to Paul) become a gentle tickle monster as she squirms in my arms, giggling in relief and insisting, *Do it again. Be the fiery dragon.*

＝

The deeper into the earth you go, the hotter it gets. So far we've barely scraped the surface in tapping geothermal energy. It is not easy to approach the heart: conventional drilling tools melt, lasers are inefficient since lots of their energy is absorbed by dust. A team in Russia made a twenty-year attempt to bore deep enough to harness heat, gave up at 40,000 feet, sealed the hole and walked away. The new hope is fusion technology—beaming millimeter-wave energy at rock and melting it without touching it, which although invasive, has a kind of respectful purity. An MIT spin-off called Quaise plans to dig down twelve miles where water, hitting rocks in 500 C heat will instantly vaporize, and the steam be converted to electricity. Quaise may use existing power plants for geothermal, leaving fossil fuels where they lie.

══

In her red ruffle dress, Ella sits in her booster seat at the kitchen table spooning blobs of yogurt into her bowl. *Can I have some jam, please?* she asks, her hazel eyes sparkling, her tangled hair in a crumpled ponytail. She calculates that good manners will prove effective. But we (and the world) haven't extinguished the fires, not yet, though with her success in getting what she wants, perhaps those fires are less potent. I don't know what to make of that. Her will and free spirit and decisiveness and energy are a force of nature, maybe the best of human nature, which should take her far and let her find those elements in worlds beyond her own. I look at her and try to imagine her decades from now, thriving on the heat in her heart and igniting others, strong as fused red stone on a mountain and protective of all that is sacred beneath her feet.

Author's Note

I am grateful to the following journals that published many of the essays in this collection.

A previous version of "Eyes in the Soles of My Feet" was first published by *KROnline* in 2019. "Fearful Asymmetry" (2019) and "Dead Man's Fingers" (2023) appeared first in *Pembroke Magazine*; "The Nimble Cuttlefish" was first published by *South Dakota Review* (2023); "Into That Good Night" was first published by *Hamilton Stone Review* (2020). "Your Ham is a Pig" appeared originally in *LA Review* (2022); "Down a Dark Hole" appeared first in *Terrain.org.* (2023). *North American Review* first published "White Ash" (2024); *Southern Humanities Review* first published "Ghost Plant" (2022) and "Eclipsed" (2012). "Wind" (2019), a previous version of "Beating it to the Sea," (2019), and a previous version of "Fin" (2013) appeared first in *Ascent*. "Life Inside" was originally published by *Cimarron Review* in 2020. "Perfect Kill" was first published by *Gulf Coast* in 2019. A previous version of "Criminals We Know" (2019) first appeared in *Green Mountains Review*.

Acknowledgments

I'M DEEPLY GRATEFUL TO my publisher and editor Tim Schaffner, whose literary insight and advice, sense of humor and responsiveness were invaluable. Many thanks to my publicist Scott Manning for his enthusiasm, expertise, and creative approaches to marketing. I'd like to thank Michael Metivier for his thoughtful editing with a focus on scientific elements in the essays. For the design and cover art, I'm grateful to Jordan Wannemacher, who skillfully captured the spirit of the book.

Warm thanks go to my close friends who read these essays over the past seven years: Alina Shevlak, Dawn Watson, Lindsay DuPont, Rik Kirkland, Ann Hyman, Christine Matthäi, and Adrian Eisenhower—your interest and enthusiasm were an inspiration. To my family, Andy Sutton, Paul and Suzanne Dumaine, Sophia Dumaine and Alex Meyers, thank you for listening to the travails of publishing and offering positive support.

Heartfelt thanks go to my husband, Brian Dumaine—you are a saint, daily weathering my questions with patience and equanimity. Your editorial expertise was so helpful, especially your keen eye for organization, decisiveness, and advice on what to slash and burn.

Finally, my gratitude and love go to Ella for finding beauty where I had not, for seeing the little creatures I would have missed, for sharing her curiosity, and for making me keep my eyes open.

Sources

WATER

Eyes in the Soles of My Feet

Burkeman, Oliver. "Why Can't the World's Greatest Minds Solve the Problem of Consciousness?" *The Guardian*, January 21, 2015. https://www.theguardian.com/science/2015/jan/21/-sp-why-cant-worlds-greatest-minds-solve-mystery-consciousness/.

Chesler, Caren. "Medical Labs May Be Killing Horseshoe Crabs." *Scientific American*, June 9, 2016. https://www.scientificamerican.com/article/medical-labs-may-be-killing-horseshoe-crabs/.

Cramer, Deborah. "When the Horseshoe Crabs are Gone, We'll Be in Trouble." *The New York Times*, February 16, 2023. https://www.nytimes.com/2023/02/16/opinion/drug-safety-horsehoe-crab.html.

Eisner, Chiara. "Coastal Biolabs are Testing More Horseshoe Crabs with Little Accountability." *NPR*, June 30, 2023. https://www.npr.org/2023/06/10/

1180761446/coastal-biomedical-labs-are-bleeding-more-horseshoe-crabs-with-little-accountability.

Florida Fish and Wildlife Conservation Commission. "Facts About Horseshoe Crabs and FAQ." Access January 20. 2025. https://myfwc.com/research/saltwater/crustaceans/horseshoe-crabs/facts/

"Horseshoe Crab." *Encyclopaedia Britannica*. Last modified January 18, 2025. https://www.britannica.com/animal/horseshoe-crab

Kristoffer, Whitney, and Jolie Crunelle. "New Synthetic Horseshoe Crab Blood Could Mean Pharma Won't Bleed the Species Dry." *Smithsonian*, October 13, 2023.https://www.smithsonianmag.com/innovation/new-synthetic-horseshoe-crab-blood-could-mean-pharma-wont-bleed-the-species-dry-180983054/.

Levitan, Ben. "A Pathway to End the Medical Harvest of Horseshoe Crabs." *Earthjustice*, July 29, 2024. https://earthjustice.org/experts/ben-levitan/a-pathway-to-end-the-medical-harvest-of-horseshoe-crabs.

Madrigal, Alexis C. "The Blood Harvest." *The Atlantic*, February 26, 2014. https://www.theatlantic.com/technology/archive/2014/02/the-blood-harvest/284078/.

National Park Service, U.S. Department of the Interior. "Horseshoe Crab: A Living Fossil." Sandy Hook, Gateway National Recreation Area. https://www.nps.gov/gate/learn/nature/upload/nature_horseshoe_crab.pdf

Physicians Committee for Responsible Medicine. "Years of Advocacy Pay Off With Policy Change That Improves Science and Protects Horseshoe Crabs." July 26, 2024. https://www.pcrm.org/news/news-releases/years-advocacy-pay-policy-change-improves-science-and-protects-horseshoe-crabs.

"What We Learned About Vision—From Horseshoe Crabs." *20/20*, July 2021. https://www.2020mag.com/article/what-we-learned-about-vision-from-horseshoe-crabs.

World Wildlife Fund. "What does 'Endangered Species' Mean?" Accessed January 20, 2025. https://www.worldwildlife.org/pages/what-does-endangered-species-mean.

Fearful Asymmetry

Christy, John. "The Design of a Beautiful Weapon." *Smithsonian: Ocean*, September 2013. https://ocean.si.edu/ocean-life/invertebrates/design-beautiful-weapon.

Crew, Bec. "The Never-Before-Seen Footage Reveals the Violent Purpose of the Narwhal's Tusk." *Science Alert*, May 15, 2017. https://www.sciencealert.com/never-before-seen-behaviour-reveals-the-violent-purpose-of-the-narwhal-s-tusk.

Edmonds, Patricia. "Male Crabs Claw Their Way to Successful Seductions." *National Geographic*, August 2016. https://www.nationalgeographic.com/magazine/2016/08/basic-instincts-fiddler-crab-claw-size/.

"Mystery of 'Unicorn' Whale Solved." *Live Science*, December 15, 2005. https://www.livescience.com/3980-mystery-unicorn-whale-solved.html.

Springer. "Fighting Fiddler Crabs Call Each Other's Bluff." Phys.org, April 5, 2016. https://phys.org/news/2016-04-fiddler-crabs-bluff.html.

Swanson, B.O., M. N. George, S.P. Anderson, et al. "Evolutionary variation in the mechanics of fiddler crab claws." *BMC Evolutionary Biology* 13, no. 137 (2013). https://doi.org/10.1186/1471-2148-13-137.

Dead Man's Fingers

"Brito, Christopher. "3 Dogs Die Hours After Playing in Pond Filled with Toxic Algae." *CBS News*, August 13, 2019. https://www.cbsnews.com/news/blue-green-algae-dogs-north-carolina-wilmington-pond-poison-cynobacteria/.

Bufalino, Jamie. "Toxic Algae Was Everywhere This Summer." *East Hampton Star*, October 23, 2020. https://www.easthamptonstar.com/government/20201023/toxic-algae-was-everywhere-summer.

Meehan, Miranda, and Michelle Mostrom. "Cyanobacteria Poisoning (Blue-green Algae)." North Dakota State University, April 2021. https://www.ag.ndsu.edu/publications/livestock/cyanobacteria-poisoning-blue-green-algae.

Golofaro, Fred. "Editor's Log: DEC Update on Bunker Die Off." *The Fisherman*, May 13, 2021. https://www.thefisherman.com/article/editors-log-dec-update-on-bunker-die-off/.

LoBue, Carl. "Emerald Waters." *Fire Island and Beyond.* Accessed January 20, 2025. https://www.fireislandandbeyond.com/long-islands-water-quality-part-6-georgica-pond/.

"The Sad Fate of the Menhaden in the Hudson." *Tribeca Citizen,* July 7, 2020. https://tribecacitizen.com/2020/07/07/the-sad-fate-of-the-menhaden-in-the-hudson/.

"Scientists Split on Soviet Signals." *The New York Times,* October 7, 1957. Accessed January 20, 2025. https://archive.nytimes.com/www.nytimes.com/partners/aol/special/sputnik/sput-12.html.

Smithsonian Institution. "Codium fragile ssp. Fragile." *Smithsonian: Nemesis.* Accessed January 20, 2025. https://invasions.si.edu/nemesis/species_summary/6897.

"The Soviet Union's Big Surprise." *The New York Times,* October 5, 1957. Accessed January 20, 2025. https://archive.nytimes.com/www.nytimes.com/partners/aol/special/sputnik/main.html.

Stewart Van Patten, Margaret "Peg" and Dr. Charles Yarish. "Bulletin No. 39: Seaweeds of Long Island Sound" (2009). *Bulletins.* Paper 40. http://digitalcommons.conncoll.edu/arbbulletins/40.

Walsh, Christopher. "Georgica Residents Fund Research." *East Hampton Star,* May 28, 2015. https://www.easthamptonstar.com/archive/georgica-pond-residents-fund-research.

Wilford, John Noble. "With Fear and Wonder in its Wake, Sputnik Lifted Us into the Future." *The New York Times,* September 25, 2007. https://www.nytimes.com/2007/09/25/science/space/25sput.html.

Yakas, Ben. "What the Heck is Going on With All the Dead Fish in the Waters Around NYC?" *Gothamist,* December 23, 2020. https://gothamist.com/news/what-heck-going-all-dead-fish-waters-around-nyc.

The Midnight Zone

SECTION 1

Anderson, William D. "Species of Conservation Concern. Whelks Guild." South
    Carolina Department of Natural Resources. Accessed January 20, 2025.https://
    www.dnr.sc.gov/swap/supplemental/marine/whelksguild2015.pdf.
"Channeled whelks and Knobbed whelks both inhabit Barnegat Bay." Barnegat
    Bay Shellfish LLC, 2019. Accessed January 20, 2025. https://barnegatshellfish.
    org/whelk01.htm.

SECTION 2

Hickey, Hannah. "Bowhead Whales, the 'Jazz Musicians' of the Arctic, Sing
    Many Different Songs." UWNews, April 3, 2018.http://www.washington.
    edu/news/2018/04/03/bowhead-whales-the-jazz-musicians-of-the-arctic-
    sing-many-different-songs/
"19th-Century Harpoon Gives Clue on Whales." *The New York Times,* June 13, 2007.
    https://www.nytimes.com/2007/06/13/world/americas/13iht-whale.1.6123654.html.
Quackenbush, Lori. "Bowhead Whale." Alaska Department of Fish & Game,
    2008.https://www.adfg.alaska.gov/static/education/wns/bowhead_whale.pdf.
Stafford, K.M., C. Lydersen, Ø. Wiig, et al. "Extreme diversity in the songs of
    Spitsbergen bowhead whales." The Royal Society. Biology Letters 14:20180056.
    http://doi.org/10.1098/rsbl.2018.0056.

SECTION 3

Demetre, D. C. "Can Jellyfish Evaporate in the Sun?" *Natureweb*, September 27,
    2022.https://natureweb.co/can-jellyfish-evaporate/.
"The jellyfish that never dies." *BBC Earth.* Accessed January 20, 2025. https://www.
    bbcearth.com/news/the-jellyfish-that-never-dies.
"Jellyfish and Comb Jellies." *Smithsonian: Ocean.* Accessed January 20, 2025.
    https://ocean.si.edu/ocean-life/invertebrates/jellyfish-and-comb-jellies.
Orlando Science Center. "Are Jellyfish Older Than Dinosaurs?" May 27, 2021.
    https://www.osc.org/are-jellyfish-older-than-dinosaurs/.

Oskin, Becky. "First Jellyfish Genome Reveals Ancient Beginnings of Complex Body Plan." UC San Diego, December 3, 2018. https://today.ucsd.edu/story/first_jellyfish_genome_reveals_ancient_beginnings_of_complex_body_plan.

Osterloff, Emily. "Immortal Jellyfish: The Secret to Cheating Death." Natural History Museum, London. Accessed January 20, 2025. https://www.nhm.ac.uk/discover/immortal-jellyfish-secret-to-cheating-death.html.

Peach, Gary. "Wave of Jellyfish Clogs up Swedish nuclear reactor, shuts it down." NBC News, October 1, 2013. https://www.nbcnews.com/sciencemain/wave-jellyfish-clogs-swedish-nuclear-reactor-shuts-it-down-8c11311141.

SECTION 4

"Ctenophore: Scientists Identify Jellyfish-Like Animal as the Oldest Living Creature on Earth." *BBC*, May 22, 2023. https://www.bbc.co.uk/newsround/65636660.

Osborne, Mary. "Comb Jellies May Be the World's Oldest Animal Group." *Smithsonian Magazine,* May 18, 2023. https://www.smithsonianmag.com/smart-news/comb-jellies-may-be-the-worlds-oldest-animal-group-180982209/.

Ralls, Eric. "Oldest Living Creature on Earth Identified, emerging 700 million years ago." Earth.com, May 17, 2023. https://www.earth.com/news/oldest-living-creature-on-earth-identified-emerging-700-million-years-ago/.

Sanders, Robert. "What Did the Earliest Animals Look Like?" *UC Berkeley News,* May 17, 2023. https://news.berkeley.edu/2023/05/17/what-did-the-earliest-animals-look-like/.

Schultz, D.T., S.H.D. Haddock, J.V. Bredeson, et al. "Ancient gene linkages support ctenophores as sister to other animals." *Nature* 618, 110–117 (2023). https://doi.org/10.1038/s41586-023-05936-6.

Wright, Jeremy. "Ctenophora Comb Jellies." *Animal Diversity Web* (Museum of Zoology, University of Michigan), 2014. Accessed September 10, 2024. https://animaldiversity.org/accounts/Ctenophora/.

SECTION 5

Cox, David. "Can This Weird Shark Show Us How to Live for Centuries?" NBC News, October 23, 2017. https://www.nbcnews.com/mach/science/can-weird-shark-show-us-how-live-centuries-ncna812671.

Morrelle, Rebecca. "400-year-old Greenland Shark 'Longest Living Vertebrate.'" *BBC*, August 12, 2016. https://www.bbc.com/news/science-environment-37047168.

Nielsen, Julius, Rasmus B. Hedeholm, Jan Heinemeier, et al. "Eye Lens Radiocarbon Reveals Centuries of Longevity in the Greenland Shark *(Somniosus microcephalus)*." *Science* 353 (6300), August 12, 2016. https://www.science.org/doi/10.1126/science.aaf1703.

SECTION 6

Cohen, Michael P. "Oldest Living Tree Tells All." *Terrain.org*, Winter/Spring 2004. https://www.terrain.org/essays/14/cohen.htm.

Littin, Shelley. "Keepers of Prometheus: The World's Oldest Tree." Shamskm.com, January 25, 2013. https://www.shamskm.com/keepers-of-prometheus-the-worlds-oldest-tree/.

McKinnon, Shaun. "The World's Oldest Tree Might or Might Not be Sitting in a Warehouse in Tuscon." AZCentral, October 3, 2015. https://www.azcentral.com/story/news/local/best-reads/2015/10/04/worlds-oldest-tree-tucson-warehouse-possibility-prometheus-tree/73211730/.

Miranda, Carolina A. "Follow Up—More Tales of the Prometheus Tree and How It Died." *The Los Angeles Times*, February 28, 2015. https://www.latimes.com/entertainment/arts/miranda/la-et-cam-video-prometheus-bristlecone-pine-20150227-column.html.

Possession

Bilger, Burkhard. "Swamp Things." *The New Yorker*, April 20, 2009. https://www.newyorker.com/magazine/2009/04/20/swamp-things.

Bishop, Sydney. "Florida's Python Challenge is just wrangling snakes for some. But for military vets, it's a chance to heal." *CNN*, August 12, 2024. https://

www.cnn.com/2024/08/10/us/florida-everglades-python-challenge-veterans/
index.html.

Bui, Lily. "A Chemical Attack that Killed a Countryside & Scarred a People."
Nautilus, July 20, 2015. https://nautil.us/a-chemical-attack-that-killed-a-
countryside-scarred-a-people-235553/.

Esch, Mary. "Pine-Killing Southern Beetle May Be More Deadly in North."
Phys.org, June 10, 2018. https://phys.org/news/2018-06-pine-killing-southern-
beetle-deadly-north.html.

Janos, Adam. "How Burmese Pythons Took Over the Florida Everglades." *History.
com*, February 20, 2020, updated August 11, 2023. https://www.history.com/
news/burmese-python-invasion-florida-everglades.

Matat, Staphany. "A Hunter's Graveyard Shift: Grabbing Pythons in the
Everglades." *ABC News*, August 18, 2024. https://abcnews.go.com/US/
wireStoMry/hunters-graveyard-shift-grabbing-pythons-everglades-112924805.

Swenson, Kyle. "Military Vets Heal PTSD by Capturing Burmese Pythons in
the Everglades." *The Miami New Times*, December 9, 2014. https://www.
miaminewtimes.com/news/military-vets-heal-ptsd-by-capturing-burmese-
pythons-in-the-everglades-6552384 .

United States Geological Survey. "FAQ: How Have Invasive Pythons Impacted
Florida Ecosystems?" Updated July 20, 2024. https://www.usgs.gov/faqs/
how-have-invasive-pythons-impacted-florida-ecosystems.

Dependence by Design

Dunn, Casey. "Siphonophores." *Current Biology* 19 (6), March 24, 2009. https://
www.cell.com/current-biology/fulltext/S0960-9822(09)00675-7.

King, Rachel. "The Portuguese man-of-war (*Physalia physalis*)" Southeastern
Regional Taxonomic Center. South Carolina Department of Natural
Resources. Accessed January 20, 2025. https://www.dnr.sc.gov/marine/sertc/
The%20Portuguese%20man.pdf.

Kurlansky, Mindy B. *"Physalia physalis." Animal Diversity Web* (Museum of

Zoology, University of Michigan), 2002. Accessed September 10, 2024. https://animaldiversity.org/accounts/Physalia_physalis/.

Munro, Catriona, Zer Vue, Richard R. Behringer, Casey W. Dunn. "Morphology and Development of the Portuguese man of war, *Physalia physalis.*" *Scientific Reports* 9 (15522), October 29, 2019. https://www.ncbi.nlm.nih.gov/pmc/articles/PMC6820529/.

Beating it to the Sea

Bennett, Larisa. "Sea Turtles: Cheloniidae and Dermatochelyidae." *Smithsonian: Ocean,* December 2018. https://ocean.si.edu/ocean-life/reptiles/sea-turtles.

Center for Biological Diversity. "Loggerhead Sea Turtle." Accessed January 20, 2025.https://www.biologicaldiversity.org/species/reptiles/loggerhead_sea_turtle/natural_history.html.

Florida Fish and Wildlife Conservation Commission. "The History and Life of a Sea Turtle." Accessed January 20, 2025. https://myfwc.com/research/wildlife/sea-turtles/florida/life-history.

"Folklore and Sea Turtles," *Marine Turtle Newsletter* 61 (29), 1993. Accessed January 20, 2025. http://www.seaturtle.org/mtn/archives/mtn61/mtn61p29.shtml.

Handy, E. S. C. and E. G. Handy with collaboration of Mary Kawena Pukui. *Native Planters in Old Hawaii—Their Life, Lore, and Environment.* Bishop Museum Press, 1972.

Oceana. "Feature: Migration of Sea Turtles." https://europe.oceana.org/migration-of-sea-turtles/#:~:text=At%20the%20water's%20edge%2C%20hundreds,across%20the%20entire%20North%20Atlantic.

National Wildlife Federation. "Sea Turtles." Accessed January 20, 2025. https://www.nwf.org/Educational-Resources/Wildlife-Guide/Reptiles/Sea-Turtles.

Yee, Tammy. "The Legend of Kauila at Punalu'u." *Turtle Talk.* Accessed January 20, 2025.https://www.tammyyee.com/tt-kauila.html.

Whiteman, Lily. "Loggerhead Turtle Migration Follows Magnetic Map." *Live Science,* June 20, 2012. https://www.livescience.com/21080-loggerhead-turtle-migration.html.

The Nimble Cuttlefish

Bates, Mary. "Cooperative Fish Take Turns with Gender Roles." *Phys.Org,* May 13,
 2016.https://phys.org/news/2016-05-cooperative-fish-gender-roles.html.

British Sea Fishing. "Additional Deep Sea Species." Accessed January 20, 2025.
 http://britishseafishing.co.uk/additional-deep-sea-species/.

Godfrey-Smith, Peter. *Other Minds: The Octopus, the Sea, and the Deep Origins of
 Consciousness.* Farrar, Straus and Giroux, 2016.

Ho, Leonard. "Hermaphroditism: A Tale of Two Sexes." *Reefscapes.net,* 2002.
 Accessed January 20, 2025. http://www.reefscapes.net/articles/articles/2002/
 hermaphroditism.html.

Kelly, Erin. "10 Sex-Changing Animals That Don't Adhere to Gender
 Rolls." *All That's Interesting,* July 16, 2016. http://allthatsinteresting.com/
 sequential-hermaphrodotism-sex-changing-animals.

Klinghoffer, David. "Chalk Bass, a Fish for Our Time, Changes Gender Roles
 20 Times a Day." *Evolution News,* July 8, 2016. https://evolutionnews.
 org/2016/07/chalk_bass_a_fi/.

Marranzino, Ashley. "Sea of Love: Hermaphroditic Fishes." *Oceanbites,* February
 11, 2016.https://oceanbites.org/sea-of-love-hermaphroditic-fishes/.

Pappas, Stephanie. "Tricky Cuttlefish Put on Gender-Bending Disguise." *Live
 Science,* July 3, 2012. https://www.livescience.com/21374-cuttlefish-gender-
 bending-disguise.html.

Sidder, Aaron. "These Fish Swap Their Sex Up to 20 Times a Day." *National
 Geographic,* July 6, 2016. https://news.nationalgeographic.com/2016/07/
 sex-swapping-fish-chalk-bass-hermaphrodites/.

Woolf, Virginia. *Orlando: A Biography.* Hogarth Press, 1928.

Into That Good Night

Bodkin, Henry. "Power Naps and Eating on the Wing—How Common Swifts
 Set 10 Month Flight Record en route from Britain to Southern Africa." *The
 Telegraph,* October 27, 2016. https://www.telegraph.co.uk/science/2016/10/27/
 common-swift-sets-new-record-by-staying-airborne-for-10-months-a/.

Gannon, Megan. "Can Any Animal Survive Without Sleep?" *LiveScience*, March 2, 2019.https://www.livescience.com/64873-can-animals-survive-without-sleep.html.

Hecker, Bruce. "How Do Whales and Dolphins Sleep Without Drowning?" *Scientific American*, February 2, 1998. https://www.scientificamerican.com/article/how-do-whales-and-dolphin/.

Homer, *The Odyssey*. Translated by Robert Fagles. Penguin, 1999.

Kenya Wildlife Safaris. "Sleeping Patterns of Giraffes in Kenya." Accessed January 20, 2025. https://www.safari-center.com/sleeping-patterns-of-giraffes-in-kenya/.

Landy, Evan. "Everything You Ever Wanted to Know about the Common Swift." *Animalogic*, June 14, 2016. https://blog.animalogic.ca/blog/everything-you-ever-wanted-to-know-about-the-common-swift.

O'Connell, Lindsey. "Who Needs Sleep Anyway?" *Ask a Biologist* (Arizona State University). Accessed January 20, 2025. https://askabiologist.asu.edu/plosable/who-needs-sleep-anyway.

Suni, Eric. "How Do Animals Sleep?" Sleepfoundation.org, December 21, 2023. https://www.sleepfoundation.org/animals-and-sleep.

Program Notes

Chavez, Nicole. "Truck Driver Plows into Peru's 2,000-Year-Old Archeological Enigma."CNN, February 1, 2018. https://www.cnn.com/2018/02/01/americas/nazca-lines-peru-truck-driver/index.html.

"Nazca Lines." History.com, Updated August 21, 2018. https://www.history.com/topics/south-america/nazca-lines.

EARTH

Down a Dark Hole

Beaulieu, David. "How to Get Rid of Voles in the Yard." *The Spruce,* July 23, 2024. https://www.thespruce.com/vole-control-getting-rid-of-voles-2131148.

Tucker, Abigail. "What Can Rodents Tell Us About Why Humans Love?" *Smithsonian,* February 2014. https://www.smithsonianmag.com/

science-nature/what-can-rodents-tell-us-about-why-humans-love-180949441.

Yong, Ed. "Consoling Voles Hint at Animal Empathy." *The Atlantic,*
January 21, 2016.https://www.theatlantic.com/science/archive/2016/01/
consoling-voles-reignite-debate-about-animal-empathy/425034/.

Chemical Warfare

Emerson, Ralph Waldo. "The Fortune of the Republic,"*Miscellanie*s. Houghton,
Mifflin and Company, 1904.

Emerson, Ralph Waldo. *The Journals and Miscellaneous Notebooks of Ralph Waldo
Emerson, Volume IV: 1832-1834,* edited by Alfred R. Ferguson. The Belknap
Press of Harvard University Press, 1964.

Gift, Nancy. *Good Weed Bad Weed: Who's Who, What to Do, and Wy Some Deserve a
Second Chance.* St. Lynn's Press, 2011.

Jameson, Sarah. "Lawn Care & Landscaping Industry Statistics."*Lawn Chick,*
October 17, 2023. https://lawnchick.com/lawn-care-statistics/.

"Market size of landscaping services in the United States From 2013-2023."
Statista.com. Accessed January 20, 2025. https://www.statista.com/
statistics/294212/revenue-of-landscaping-services-in-the-us/.

Penn State Extension. "Weed Identification and Control." Accessed January 20, 2025.
https://extension.psu.edu/introduction-to-weeds-what-are-weeds-and-why-
do-we-care.

Richardson Jr., *Robert D. Emerson: The Mind on Fire.* University of California
Press, 1995.

Roberts, Catherine. "A Lush Lawn Without Pesticides." *Consumer Reports,*
April 8, 2021.https://www.consumerreports.org/lawn-care/a-lush-lawn-
without-pesticides-a3940080625/.

Tilly, Nikki. "What is a Weed: Weed Info and Control Methods in Gardens."
Gardening, July 6, 2021. https://www.gardeningknowhow.com/plant-problems/
weeds/what-is-a-weed.htm.

University of Minnesota Extension. "Is This Plant a Weed?" Accessed January 20,
2025. http://www.extension.umn.edu/garden/diagnose/weed/.

White Ash

Albeck-Ripka, Livia. "Saving the Fire Victims Who Cannot Flee: Australia's Koalas."
   The New York Times, November 14, 2019, updated October 30, 2023. https://www.
   nytimes.com/2019/11/14/world/australia/australia-koalas-fire.html.

Congressional Research Service. "Wildfire Statistics." *In Focus*. Updated June 1,
   2023. https://fas.org/sgp/crs/misc/IF10244.pdf.

Frazer, Jennifer. "Dying Trees Can Send Food to Neighbors of Different Species."
   Scientific American, May 9, 2015. https://blogs.scientificamerican.com/
   artful-amoeba/dying-trees-can-send-food-to-neighbors-of-different-species/.

Struzik, Ed. "Does a Fire-Ravaged Forest Need Human Help to Recover?"
   Yale Environment360, June 14, 2018. https://e360.yale.edu/features/
   does-a-fire-ravaged-forest-need-human-help-to-recover.

Weiss, Miranda. "Trees on the Move." *American Forests*, June 11, 2019. https://www.
   americanforests.org/magazine/article/trees-on-the-move/.

Wuerthner, George. "The Ecological Value of Dead Trees." *The Wildlife
   News*, December 18, 2018. https://www.thewildlifenews.com/2018/12/20/
   the-ecological-value-of-dead-trees/.

Facetime with a Mole

Reuell, Peter. "Looking at Lunglessness." *The Harvard Gazette*,
   January 31, 2019. https://news.harvard.edu/gazette/story/2019/01/
   lungless-salamanders-skin-expresses-protein-crucial-for-lung-function/.

Smithsonian's National Zoo & Conservation Biology Institute. "Eastern Red-
   Backed Salamander." Accessed January 20, 2025. https://nationalzoo.si.edu/
   animals/eastern-red-backed-salamander.

University of Zurich. "How the Mole Got Its 12 Fingers." *Phys.org*, July 14, 2011.
   https://phys.org/news/2011-07-mole-fingers.html.

Like a Sycamore, Like a Laurel, Like a Dove

Emerson, Ralph Waldo. *Nature and Selected Essays*. Penguin Books, 2003.

Fazio, James R. "London Planetree: A Tough Dweller with a Confusing Past." Arbor Day Foundation, December 5, 2017. https://arbordayblog.org/ treeoftheweek/london-planetree-tough-dweller-confusing-past/.

H. David. "London Planes and American Sycamores." *Growing History* (The Philadelphia Historic Plants Consortium), March 13, 2012. https://growinghistory.wordpress.com/2012/03/13/ london-planes-and-american-sycamores/.

Nix, Steve. "Sycamore—Not Just a Planetree." *Treehugger*, April 14, 2017. https:// www.thoughtco.com/sycamore-overview-and-lore-1343160

Ovid, *Metamorphoses*. Translated by A. S. Kline. Borders Classics, 2004.

Poems of the Masters: China's Classic Anthology of T'Ang and Sung Dynasty Verse. Translated by Red Pine. Copper Canyon Press, 2003.

Sheridan, Moira. "Behold, the Beautiful but Problematic Sycamore Tree." *The News Journal* (Wilmington, Delaware), December 15, 2015. https://www.delawareonline. com/story/life/2015/12/15/beautiful-but-problematic-sycamore/77376772/.

Stevens, Wallace. "*An Ordinary Evening in New Haven,*" *The Collected Poems of Wallace Stevens*. Alfred A. Knopf, 1976.

Still, Douglas, and Fiona Watt. "Why the Sycamore Sheds its Bark." *The Daily Plant* (New York City Parks), Volume XIX, October 14, 2004. https://www. nycgovparks.org/news/daily-plant?id=19242.

Taft, Dave. "London Plane: A Tree With Gritty Roots." *The New York Times*, December 22, 2016. https://www.nytimes.com/2016/12/22/nyregion/london-plane-tree.html.

Williams, David R. "The Moon Trees." NASA Goddard Space Flight Center. Updated October 28, 2024. https://nssdc.gsfc.nasa.gov/planetary/lunar/moon_ tree.html.

Ghost Plant

Overbye, Dennis. "Darkness Visible, Finally: Astronomers Capture First Ever Image of a Black Hole." *The New York Times*, April 10, 2019. https://www. nytimes.com/2019/04/10/science/black-hole-picture.html.

AIR

Wind

Gabert, Bill. "Officials Investigating the Roles of Wind and Power Lines in Northern California Wildfires." *Wildfire Today*, October 17, 2017. http:// wildfiretoday.com/2017/10/17/officials-investigating-the-roles-of-wind-and-power-lines-in-northern-california-wildfires/.

Jenner, Lynn. "California's Mendocino Complex of Fires Now Largest in State's History." *Phys.org*, August 7, 2018. https://phys.org/news/2018-08-california-mendocino-complex-largest-state.html#jCp.

Mak, Aaron. "Strong Winds This Weekend Could Make the California Wildfires Even Worse." *Slate*, August 3, 2018. https://slate.com/technology/2018/08/carr-redding-fire-california-strong-winds-forecast-mendocino-complex-blaze.html.

Life Inside

Calver, Dr. Bill. "Monarch Butterflies." *Journey North* (University of Wisconsin, Madison). Accessed January 20, 2025. https://journeynorth.org/tm/monarch/Predation.html.

Gomez, Tony. "13 Monarch Predators in the Butterfly Garden." *Monarch Butterfly Garden*. Accessed January 20, 2025. https://monarchbutterflygarden.net/stop-monarch-predators/.

Lohmiller, George and Becky. "Common Milkweed: Uses and Natural Remedies." *The Old Farmer's Almanac*, May 19, 2020. https://www.almanac.com/content/common-milkweed-uses-and-natural-remedies.

National Wildlife Federation. "Monarch Butterfly." Accessed January 20, 2025. https://www.nwf.org/Educational-Resources/Wildlife-Guide/Invertebrates/ Monarch-Butterfly.

Perfect Kill

Hadley, Debbie. "10 Fascinating Facts about Dragonflies." *Thought Co.*, February 9, 2020.https://www.thoughtco.com/fascinating-facts-about-dragonflies-1968249.

Handley, Andrew. "10 Surprisingly Brutal Facts about Dragonflies." *Listverse*, April 18, 2013.https://listverse.com/2013/04/18/10-surprisingly-brutal-facts-about-dragonflies/.

Lohmiller, George and Becky. "Dragonflies: Facts, Symbolic Meaning, and Habitat." *The Old Farmer's Almanac*, July 24, 2019. https://www.almanac.com/ content/dragonflies-facts-symbolic-meaning-and-habitat.

"Meganeura." *A-Z-animals.com*. Accessed January 21, 2025. https://a-z-animals. com/animals/meganeura.

Zielinski, Sarah. "14 Fun Facts About Dragonflies." *Smithsonian*, October 5, 2011. https://www.smithsonianmag.com/science-nature/14-fun-facts-about-dragonflies-96882693/.

Criminals We Know

Awford, Jenny. "Why We All Think Seagulls are Foul: Birds are Most Hated Because of Way They Steal Food and Cover Cars in Mess." *DailyMail.co.uk*, October 12, 2014. https://www.dailymail.co.uk/news/article-2790229/why-think-seagulls-foul-birds-hated-way-steal-food-cover-cars-mess.html.

Center for Coastal Studies (Provincetown, Massachusetts). "Gulls and Terns." Accessed January 20, 2025. https://coastalstudies.org/connect-learn/ stellwagen-bank-national-marine-sanctuary/sea-birds/gulls-and-terns/.

Cornell Lab of Ornithology, *All About Birds*. https://www.allaboutbirds.org/news/.

Ellis, Peter Berresford. *A Dictionary of Irish Mythology*. Oxford University Press, 1992.

"Great Black-backed Gull." *Audubon.org*. Accessed January 20, 2025. https://www.audubon.org/field-guide/bird/great-black-backed-gull.

"Herring Gull." *Audubon.org*. Accessed January 20, 2025. https://www.audubon.org/field-guide/bird/herring-gull.

Montgomery, Sy. "Gull." *Encyclopaedia Britannica*. Updated January 20, 2025. https://www.britannica.com/animal/gull.

"Why Do Gulls Perform a Rain Dance?" *Bird Spot*. Access January 20, 2025. https://www.birdspot.co.uk/bird-brain/why-do-gulls-perform-a-rain-dance.

Yong, Ed. *An Immense World*. Random House, 2022.

Brian and the Snowy Egrets

Connecticut Department of Energy and Conservation. "Snowy Egret." Accessed January 20, 2025. https://portal.ct.gov/deep/wildlife/fact-sheets/snowy-egret.

"Snowy Egret." *All About Birds* (Cornell Lab of Ornithology). Accessed January 20, 2025. https://www.allaboutbirds.org/guide/Snowy_Egret.

"Snowy Egret." *Audubon.org*. Accessed January 20, 2025. https://www.audubon.org/field-guide/bird/snowy-egret.

Stung

University of Maryland Extension. "Social Wasps: Yellowjackets, Hornets, and Paper Wasps." Updated October 17, 2024. https://extension.umd.edu/resource/social-wasps-yellowjackets-hornets-and-paper-wasps/.

FIRE

Speaking Light

"Bioluminescence Questions and Answers." Latz Laboratory, Scripps Institution of Oceanography, UC San Diego. Accessed January 20, 2025. https://latzlab.ucsd.edu/bioluminescence/bioluminescence-questions-and-answers.

"The Black Dragonfish." The Bioluminescence Web Page (University of California,

Santa Barbara), September 1, 1999. https://biolum.eemb.ucsb.edu/pdf/dragon.pdf.

Kandinsky, Wassily. *Concerning the Spiritual in Art*. Dover Publications, 1977.

Knight, Kathryn. "Luciferin Diet Fuels Brittle Star Glow." *Journal of Experimental Biology* 223 (4), February 18, 2020. https://doi.org/10.1242/jeb.222539.

McFall-Ngai, Margaret. "Hawaiian Bobtail Squid." Department of Medical Microbiology and Immunology (University of Wisconsin). https://www.cell.com/current-biology/pdf/S0960-9822(08)01137-8.pdf.

The Ocean Portal Team. "Bioluminescence." *Smithsonian: Ocean*. April 2018. https://ocean.si.edu/ocean-life/fish/bioluminescence.

Sea and Sky. "Creatures of the Deep Sea." Accessed January 20, 2025. http://www.seasky.org/deep-sea/dragonfish.html.

Stephens, Tim. "A Little Squid and its Glowing Bacteria Yield New Clues to Symbiotic Relationships." *UC Santa Cruz News*, March 9, 2021. https://news.ucsc.edu/2021/03/bioluminescent-squid.html.

Miyashiro, Tim, and Edward G. Ruby. "Shedding Light on Bioluminescence Regulation in Vibrio fisheri." *Molecular Microbiology* 84 (5), 2012. https://doi.org/10.1111/j.1365-2958.2012.08065.x.

Sunshine

Amos, Jonathan. "Sun's Surface Seen in New Detail." *BBC News*, January 30, 2020. https://www.bbc.com/news/science-environment-51305216.

Butler, Grant. "Solar Eclipses in History: A Sense of History Turns to Awe in Science." *The Oregonian*, August 2, 2017. https://www.oregonlive.com/eclipse/2017/08/solar_eclipses_in_history_a_se.html.

Crockett, Christopher. "What Are White Dwarf Stars?" *Earth Sky*, September 18, 2020. https://earthsky.org/astronomy-essentials/white-dwarfs-are-the-cores-of-dead-stars/.

Dobrijevic, Daisy. "Coronal Mass Ejections: What Are They and How Do They Form?" *Space.com*, June 24, 2022. https://www.space.com/coronal-mass-ejections-cme.

Ferdowski, Samir. "No, Congressman, You Can't Alter the Moon's Orbit to

Stop Climate Change." *Vice,* June 9, 2021. https://www.vice.com/en/article/
louie-gohmert-asks-about-altering-earth-orbit-stop-climate-change/.

Grossman, Lisa. "The Parker Solar Probe Will Have Company on its Next Pass by
the Sun." *ScienceNews,* January 15, 2021. https://www.sciencenews.org/article/
nasa-parker-solar-probe-sun-next-close-pass-esa-orbiter.

Hall, Shannon. "Give Thanks for the Winter Solstice. You Might not be Here
Without It." *The New York Times,* updated December 21, 2022. https://www.
nytimes.com/2017/12/20/science/winter-solstice-december-21.html.

Maag, Christopher. "A Divided America Agrees on One Thing: The Eclipse
was Awesome." *The New York Times,* April 9, 2024. https://www.nytimes.
com/2024/04/09/nyregion/total-solar-eclipse.html.

"Nanoflares in the Sun's Plasma May Cause Its Scalding Atmosphere." *New
Scientist.* October 9, 2017. https://www.newscientist.com/article/2149671-
nanoflares-in-the-suns-plasma-may-cause-its-scalding-atmosphere/.

NASA. "Eclipse 101." Accessed January 20, 2025. https://eclipse2017.nasa.gov/
eclipse-history.

NASA. "Our Sun: Facts." Accessed January 20, 2025. https://solarsystem.nasa.gov/
solar-system/sun/in-depth/.

NASA. "NASA's Parker Solar Probe Sheds New Light on the Sun."
December 4, 2019. https://www.nasa.gov/feature/goddard/2019/
nasas-parker-solar-probe-sheds-new-light-on-the-sun.

Plait, Phil. "How Do Stars Really Die?" *Scientific American, A*ugust 2, 2024.
https://www.scientificamerican.com/article/how-do-stars-really-die/.

Pultarova, Tereza. "Discovery of Slow Waves on the Sun Could Shed Light on
Magnetic Field Mystery." *Space.com,* July 29, 2021. https://www.space.com/
sun-scientists-discover-new-plasma-wave.

Sharp, Tim. "The Sun's Atmosphere: Photosphere, Chromosphere and Corona." *Space.
com,* November 1, 2017. https://www.space.com/17160-sun-atmosphere.html.

Siegel, Ethan. "This is Why Einstein Knew That Gravity Must Bend Light." *Forbes,*
April 26, 2019. https://www.forbes.com/sites/startswithabang/2019/04/26/
this-is-why-einstein-knew-that-gravity-must-bend-light/?sh=51d2c5472ef6.

Siegel, Ethan. "Ask Ethan. Will the Earth Eventually be Swallowed by the Sun?" *Forbes*, February 8, 2020. https://www.forbes.com/sites/startswithabang/2020/02/08/ask-ethan-will-the-earth-eventually-be-swallowed-by-the-sun/?sh=5b3249445cb0.

University of Gottingen. "How the Sun's Magnetic Forces Arrange Gas Particles." *ScienceDaily*, October 13, 2021. https://www.sciencedaily.com/releases/2021/10/211013114104.htm.

Weisberger, Mindy. "What Were the First Records of Solar Eclipses?" *LiveScience*, July 5, 2017. https://www.livescience.com/59686-first-records-solar-eclipses.html.

"Why There is Day and Night." *Solar Folklore and Storytelling*, compiled by Deborah Scherrer. Stanford Solar Center (Stanford University). Accessed January 20, 2025. http://solar-center.stanford.edu/folklore/Solar-Folklore.pdf.

Williams, Matt. "Will Earth Survive When the Sun Becomes a Red Giant?" *Phys.org*, May 10, 2016. https://phys.org/news/2016-05-earth-survive-sun-red-giant.html.

Fiery Dragon

Bearteaux, Danielle. "More Fires, More Problems." *EOS*, February 1, 2022. https://eos.org/articles/more-fires-more-problems.

Blain, Los. "Fusion Tech is Set to Unlock Near-Limitless Ultra-Deep Geothermal Energy." *New Atlas*, February 25, 2022. https://newatlas.com/energy/quaise-deep-geothermal-millimeter-wave-drill/.

Blakemore, Erin. "This Mine Has Been Burning for Over 50 Years." *History.com*, August 3, 2023. https://www.history.com/news/mine-fire-burning-more-50-years-ghost-town.

Gemini Magazine. "Energy underfoot: Bringing up heat from inside Earth." *EarthSky.org*, October 19, 2020. https://earthsky.org/human-world/energy-underfoot-bringing-up-heat-from-the-center-of-the-planet/ .

Krajick, Kevin. "Fire in the Hole." *Smithsonian*, May 2005. https://www.smithsonianmag.com/science-nature/fire-in-the-hole-77895126/.

Montana Department of Transportation. "The Red-Capped Hills of Eastern Montana." Accessed January 20, 2025. https://www.mdt.mt.gov/travinfo/docs/

roadsigns/Red-CappedHills.pdf.

Morford, Stacy. "Where Does the Earth's Heat Come From?" *The Conversation,* December 15, 2020. https://theconversation.com/where-does-the-earths-heat-come-from-151788.

"Ring of Fire." *Encyclopaedia Britannica.* Updated January 15, 2025. https://www.britannica.com/place/Ring-of-Fire.

Simon, Matt. "Wildfires are Digging Carbon Spewing Holes in the Arctic." *Wired,* January 4, 2022. https://www.wired.com/story/wildfires-are-digging-carbon-spewing-holes-in-the-arctic/.

St. John, James. "Clinker: photo caption." *Flickr.* July 30, 2015. Accessed January 20, 2025. https://www.flickr.com/photos/jsjgeology/19509533723.

Witz, Alexandra. "The Arctic is Burning like Never Before—And that's Bad News for Climate Change." *Nature,* September 10, 2020. https://www.nature.com/articles/d41586-020-02568-y.